TABLE OF CONTENTS

RISK MANAGEMENT SERIES

Primer for Design Professionals: Communicating with Owners and Managers of New Buildings on Earthquake Risk

PROVIDING PROTECTION TO PEOPLE AND BUILDINGS

FEMA

www.fema.gov

Seismic risk management tools, including new seismic engineering technology and data, are now available to assist with evaluating, predicting, and controlling financial and personal-injury losses from future damaging earthquakes. These tools have evolved as a result of scientific and engineering breakthroughs, including new earth-science knowledge about the occurrence and severity of earthquake shaking, and new engineering techniques for designing building systems and components to withstand the effects of earthquakes. As a result, design and construction professionals can now design and construct new buildings with more predictable seismic performance than ever before.

Seismic risks can be managed effectively in a number of ways, including the design and construction of better performing buildings as well as the employment of strategies that can result in risk reduction over the life of the building. Risk reduction techniques include the use of new technologies, such as seismic isolation and energy dissipation devices for both structural and non-structural systems; site selection to avoid hazards such as ground motion amplification, landslide, and liquefaction; and the use of performance-based design concepts, which enable the engineer to better estimate building capacity and seismic loading demand and to design buildings for enhanced performance (beyond that typically provided by current seismic codes). The implementation of risk reduction strategies by building owners and managers is critically important, not only for reducing the likelihood of life loss and injury, but also for reducing the potential for losses associated with earthquake damage repair and business interruption.

> The implementation of risk reduction strategies by building owners and managers is critically important, not only for reducing the likelihood of life loss and injury, but also for reducing the potential for losses associated with earthquake damage repair and business interruption.

The Federal Emergency Management Agency (FEMA) has commissioned and funded the development of this document to facilitate the process of educating building owners and managers about seismic risk management tools that can be effectively and economically employed by them during the building development phase – from site selection through design and construction – as well as the operational phase.

This document also recognizes that seismic design professionals (architects and engineers) throughout the United States have varying levels of technical knowledge and experience pertaining to the seismic design of buildings. In areas of moderate and high seismicity, the knowledge and experience is substantially greater than in areas of low seismicity. In many cases, design professionals rely extensively, if not exclusively, on local building seismic codes for specifications and instruction for incorporating seismic resistance in buildings that they design. In other cases, design professionals supplement their design experience and knowledge by using technical resource documents on seismic-design related issues prepared by professional structural engineering organizations and institutions,[1] in many cases with funding from state and federal agencies (e.g., FEMA). As a result, many design professionals are likely to have substantial knowledge about concepts and approaches for reducing seismic risk in new buildings, the special focus of this document.

Regardless of their level of knowledge and experience in seismic design, seismic design professionals are likely to have little knowledge regarding non-engineering-related strategies and options that could be employed by building owners and managers to reduce their seismic risk. This document has therefore also been written to educate the seismic design professional on these non-engineering-related risk management approaches, including risk transfer through insurance, risk reduction through earthquake response planning, and risk reduction through other non-engineering-related means.

> **This document has also been written to educate the seismic design professional on non-engineering-related risk management approaches.**

While the methods described are general in nature and apply to most building uses, the document specifically addresses six occupancy types:

- commercial office facilities,
- retail commercial facilities,
- light manufacturing facilities,
- healthcare facilities,
- local schools (kindergarten through grade 12), and
- higher education (university) facilities.

The intended audience for this document consists of those design professionals (architects and engineers) who typically work with building owners and managers in developing new building projects. The document is intended to be used in conjunction with a set of six companion FEMA-funded brochures for building owners and managers, written to encourage the use of seismic risk management tools and strategies in the design and construction of new buildings. A brochure has been pre-

---

1. Example organizations and institutions include: the American Society of Civil Engineers, the Applied Technology Council, the Building Seismic Safety Council, and the Structural Engineers Association of California.

pared for each of the six facility types identified above, and each is limited in scope and content so that it can be quickly read and easily understood by building owners and managers. The brochures identify a number of issues, many of them posed in the form of questions, that relate to seismic risk and the benefits that seismic risk management, including performance-based design, can provide to building owners and managers. Each brochure is amply illustrated with photographs, charts, and tables that demonstrate important concepts in seismic risk management and seismic design and construction.

This document and set of brochures were preceded approximately fifteen years ago with a series of FEMA documents, known as the *Seismic Considerations Series*, which were written for a broad range of professionals and stakeholders interested in and concerned about building seismic performance issues.

1.1 IMPETUS FOR UPDATING THE PRIOR DOCUMENTS IN THE FEMA *SEISMIC CONSIDERATIONS SERIES*

The initial publications in the FEMA-funded *Seismic Considerations Series*, prepared by the Building Seismic Safety Council and published in the time period, 1988-1990, provided guidance on seismic safety and design-related issues to owners, managers, and designers of selected building types. The series consisted of the following documents:

○ *Seismic Considerations, Elementary and Secondary Schools* (FEMA 149 Report)

○ *Seismic Considerations, Health Care Facilities* (FEMA 150 Report)

○ *Seismic Considerations, Hotels and Motels* (FEMA 151 Report)

○ *Seismic Considerations, Apartment Buildings* (FEMA 152 Report)

○ *Seismic Considerations, Office Buildings* (FEMA 153 Report)

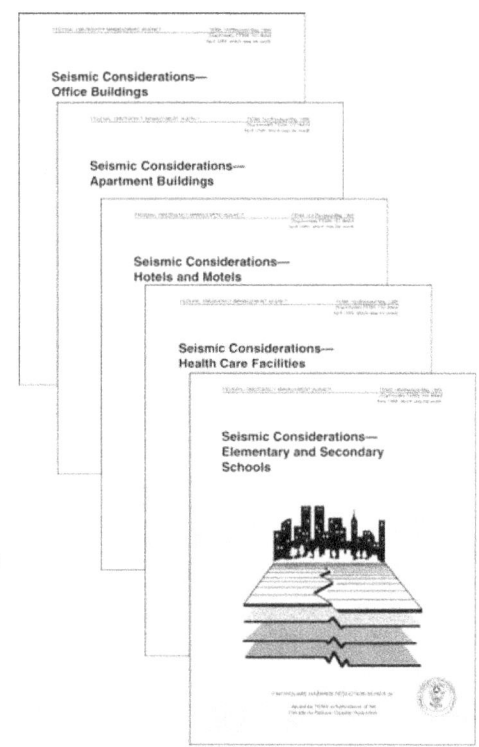

The documents were written to address seismic performance issues and cost-effective strategies for improving building seismic performance through engineering approaches and procedures laid out in the then state-of-the-art *NEHRP Recommended Provisions for the Development of Seismic Regulation for New Buildings* (BSSC, 1988).

Since 1990, a considerable amount of new knowledge and information has been developed and published under the

National Earthquake Hazards Reduction Program (NEHRP), a broad multidisciplinary and Congressionally-mandated research and development program, administered by four Federal agencies and funded at the level of approximately $100 million per year. The purpose of NEHRP is to improve the capacity of the nation's built environment to resist the effects of earthquake induced ground shaking and the collateral hazards of landslide, liquefaction, ground failure, inundation, and postearthquake fires.

Major new seismic hazard mitigation tools and strategies developed in the 1990s include:

○ new seismic hazard maps, published by the U. S. Geological Survey, that incorporate (1) state-of-the-knowledge earthquake occurrence models, (2) state-of-the-knowledge ground motion attenuation relationships, and (3) new probability-of-occurrence levels that better characterize expected ground motions in regions of large, infrequent earthquakes;

○ new performance-based seismic design concepts, criteria, and procedures, funded by FEMA and published in the FEMA 273 *NEHRP Guidelines for the Seismic Rehabilitation of Buildings* (ATC, 1997a), and its successor document, FEMA 356, *Prestandard and Commentary for Seismic Rehabilitation of Buildings* (ASCE, 2000), that enable the building owner and design engineer to evaluate and upgrade buildings to meet specific performance levels (e.g., collapse prevention, life safety, immediate occupancy, continued operation) for defined levels of earthquake ground shaking; and

○ new seismic risk management strategies, developed largely by the private sector, which enable building owners and managers to reduce the financial impacts of earthquakes by diversifying the locations of operations, by obtaining higher levels of earthquake insurance, and by using securitization instruments, such as Catastrophic Bonds.

These new technological developments provide the necessary tools for building owners and managers, with the assistance of design professionals, to make and implement cost-effective decisions regarding seismic safety and seismic hazard mitigation. They also provide the impetus and justification for updating the original *Seismic Considerations Series* documents, which are based on seismic hazard information and engineering knowledge and concepts developed in the 1970s and 1980s.

1.2 OBJECTIVES AND SCOPE OF THIS DOCUMENT

The objectives of this report are fourfold: (1) to summarize, in a qualitative fashion, important new concepts in performance-based seismic design and new knowledge about the seismic hazard facing the United States (in a way that can be easily communicated to building owners and managers); (2) to describe a variety of concepts for reducing seismic risk, including the means to reduce economic losses that are not related to engineering solutions; (3) to provide illustrative examples and graphical tools that can be used by the design community to more effectively "sell" concepts of seismic risk management and building performance improvements; and (4) to establish a means by which seismic engineering and financial risk management can be integrated to form a holistic seismic risk management plan. The overarching goal of the document is to provide a means to facilitate communications between building owners/managers and design professionals on the important issues affecting seismic risk decision making during the design and construction of new facilities, as well as the operational phase.

> The overarching goal of the document is to provide a means to facilitate communications between building owners/managers and design professionals on the important issues affecting seismic risk decision making during the design and construction of new facilities, as well as the operational phase.

Stated another way, this report may be considered as a framework for integrating seismic risk management into already well established project planning, design, and construction processes used by most owners and designers. The report is intended to be used in:

- the initial project planning stages to address siting, general building performance considerations, and how the design process can incorporate performance-based design principles;

- the budgeting phase of a project to identify the resources that can be allocated to manage risk;

- the design phase of a project to assist in the layout of structural systems, define performance objectives, and perform benefit-cost analyses of various building options; and

- the construction administration phase of a project to achieve a high level of quality assurance and control, thereby increasing the likelihood that the facility, as constructed, will perform as expected.

In addition, the report provides information that pertains to risk management strategies that are not directly part of the project planning, design, and construction processes, but that owners and managers can use to mitigate earthquakes losses. These strategies, applicable to newly constructed buildings as well as existing facilities, should be considered in conjunction with engineering design and construction strategies when developing a holistic seismic risk management plan for a new building. Thus the report is intended to also be beneficial in:

○ evaluating the benefits of earthquake insurance and quantifying coverage needs;

○ developing a postearthquake response and recovery program that may reduce down-time and potential loss of business following a major event;

○ calculating the benefits of diversifying operations geographically or among different buildings within a single campus; and

○ dealing with the risks associated with other types of hazards, both natural and man-made.

1.3 DOCUMENT CONTENTS AND ORGANIZATION

This document has been written and organized to assist building design professionals (architects and structural engineers) in communicating with building owners on earthquake risk, that is, to advise building owners on methods that could be employed to reduce their seismic risk. It is recognized that many design professionals may not be familiar with emerging concepts in (1) seismic risk management, (2) performance-based seismic design, and (3) seismic design and performance issues related to the specific occupancies discussed in this report—commercial office facilities, retail commercial facilities, light manufacturing facilities, healthcare facilities, local schools (kindergarten through grade 12), and higher education (university) facilities. These topics are therefore discussed in detail, including illustrations and tables designed to be used by the building design professionals when communicating with building owners on the means to reduce their seismic risk.

Seismic risk management is introduced and discussed in Chapter 2, including an overview discussion on seismic risk and discussions on a range of risk reduction strategies. This chapter also describes issues to be considered when developing a risk management plan, addressed in the context of the likelihood of potential losses. The identified risk reduction strategies consist of: (1) first cost or design strategies; (2) operating cost or business strategies, and (3) event response strategies. Also included are discussions on the selection of an optimal combination of risk reduction strategies, example applications of seismic risk management strategies on real buildings, and advocacy of seismic risk management.

The means for identifying and assessing earthquake-related hazards during the site selection process are described in Chapter 3. The chapter begins with a discussion of current approaches for seismic shaking hazard determination and assessment in the United States, followed by discussions on the collateral seismic hazards of surface fault rupture, soil liquefaction, soil differential compaction, landsliding, and inundation. Chapter 3 also discusses other earthquake-related hazards affecting building performance, including vulnerable transportation and utility systems (lifelines), the hazards posed by adjacent structures, the release of hazardous materials, and postearthquake fires. Specific guidance for assessing these earthquake related hazards during the site selection process, including a checklist for use by design professionals, is provided at the end of this chapter.

In Chapter 4, emerging concepts in performance-based seismic design are described. This chapter includes a discussion on (1) expected building performance when designing new buildings to current codes; and (2) state-of-the-art concepts in performance based seismic design, which were developed for the seismic rehabilitation of existing buildings and are beginning to be applied on a volunteer basis in the seismic design of new buildings. The chapter concludes with a description of next-generation performance-based seismic design products and tools for engineers and building owners/managers expected to become available over the next decade or so.

Chapter 5 focuses on ways to reduce seismic risk by improving building performance, a first cost or design risk reduction strategy. This chapter describes and discusses performance attributes of various structural systems and materials, selection of the architectural configuration, and the interaction of nonstructural components and systems with the building structure. Also included is a discussion of the costs and benefits associated with improved performance, as well as actual case studies describing structural system cost and performance considerations.

Building and expanding on the ways to improve seismic performance discussed in Chapter 5, the next six chapters (Chapters 6 through 11) briefly identify specific design issues associated with each of the six occupancy types considered in this document. In addition, each of these chapters provides examples of earthquake performance for that facility type and discusses performance expectations and requirements, and specific vulnerabilities. Chapters 6 and 7, respectively, address commercial office buildings and commercial retail buildings; Chapter 8 addresses light manufacturing facilities; Chapter 9 focuses on healthcare facilities; and Chapters 10 and 11, respectively, address local schools (kindergarten through grade 12) and higher education facilities (universities).

Chapter 12 addresses the various responsibilities of members of the design team, including the building owner, architect, structural engineer, and mechanical/electrical/plumbing engineers. This chapter also includes discussions on the added value of risk management and design and construction quality assurance.

Following Chapter 12 are a list of references and a list of individuals who participated in the development of this report.

1.4   DOCUMENT FORMATTING AND ICONS

Several icons, shown below, are used in highlighted portions of this document to emphasize pertinent information.

 The **Definition** icon defines key terms and acronyms.

 The **Case Study** icon provides practical and relevant information based on past experience.

 The **Resources** icon provides supplemental information from FEMA and other organizations that may impact design considerations and decision-making.

 The **Cost Consideration** icon identifies a value or investment cost that needs to be considered in decision-making.

 The **Risk Consideration** icon identifies a potential or real risk that needs to be considered in decision-making.

 The **Design Consideration** icon identifies a design issue that needs to be considered in decision-making.

Chapter 2 introduces and describes seismic risk management, beginning with an overview of seismic risk, followed by discussions on the holistic nature of seismic risk management and on strategies for reducing seismic risk. These strategies fall into three categories: (1) first cost or design strategies; (2) operating cost or business strategies, and (3) event response strategies. Also included in this chapter are discussions on the selection of an optimal combination of risk reduction strategies, and example applications of seismic risk management strategies, including cost and performance considerations, described in three case studies. The chapter concludes with a discussion of the importance of seismic risk management advocacy.

2.1 SEISMIC RISK: AN OVERVIEW

In general, the term "risk" is commonly used to characterize the likelihood of an unfavorable outcome or event occurring. The term "seismic risk" is used by the scientific and engineering communities to describe the likelihood of adverse consequences resulting from the occurrence of an earthquake. Seismic risk is typically defined as a function of three elements: (1) the seismic hazard or likelihood of occurrence of an earthquake and the associated severity of shaking, (2) the seismic vulnerability or expected damage to buildings and other structures given the occurrence of an earthquake, and (3) the expected consequences or losses resulting from the predicted damage. The third term, the expected consequences, is typically used to quantify the seismic risk to an individual facility, a group of facilities, or a region. For a building, these consequences or expected losses can be broadly categorized as:

Seismic Risk

Seismic Risk is typically defined as a function of three elements:

(1) the seismic hazard or likelihood of occurrence of an earthquake and the associated severity of shaking,

(2) the seismic vulnerability or expected damage to buildings and other structures given the occurrence of an earthquake, and

(3) the expected consequences or losses resulting from the predicted damage.

- ○ Casualties – the death or injury of building occupants or passersby resulting from the building collapse, blockage of exits, or failure of life safety systems;

- ○ Capital – the value of a building's structural and nonstructural systems, including the structural framing elements, partitions, cladding, and mechanical, electrical, and plumbing systems;

- ○ Contents – the value of, for example, a building's fixed and movable equipment, goods for sale, laboratory and manufacturing equipment;

○ Business Interruption – the financial cost resulting from loss of operations; this consequence can be expressed in a variety of ways, depending on the use of the facility, e.g., lost revenue, inability to treat patients, teach students or conduct research; and

○ Market Share – the future costs of losing a competitive edge; this consequence can also be expressed in a variety of ways, including loss of clients to competitors, having staff leave to work for competitors, and losing "prestige" and the business associated with an organization's reputation.

Seismic risk, as defined above, can be reduced by a reduction in any of the three elements – seismic hazard, seismic vulnerability, and expected consequences. Seismic hazard can only be reduced by relocation of the building itself, as the likelihood of an earthquake occurring at a site and the severity of shaking is a function of the regional seismicity and local geology. If the building site is a fixed variable, seismic hazard and seismic vulnerability are often considered as one factor – the likelihood that the building will sustain earthquake damage. The combination of this factor, with the expected consequences given the occurrence of the earthquake damage, results in a measure of seismic risk. Thus seismic risk can be reduced by decreasing the likelihood of building damage (e.g., by relocating the building or by increasing the earthquake resisting capacity of the structure) or by decreasing the expected consequences (e.g., by developing a response plan, geographically diversifying operations, or purchasing insurance).

The concept of seismic risk, expressed as a function of the likelihood of damage and the expected consequences, is illustrated in Figure 2-1. The likelihood of damage is shown along the horizontal axis, increasing from left to right. As mentioned above, the likelihood of damage is a function of the seismic hazard level (expected earthquake occurrences and severity of ground shaking) and the seismic vulnerability (earthquake resisting capacity of the building). The consequences or losses resulting from the earthquake damage (or "consequence" in the more general risk term) are depicted on the vertical axis, increasing from bottom to top. The quantification of seismic risk is not a simple task; however, the graph shown in Figure 2-1 is simplified qualitatively as four distinct quadrants, each of which is described below with example scenarios.

Quadrant I, Low Risk: low likelihood of damage and low consequences; examples include:

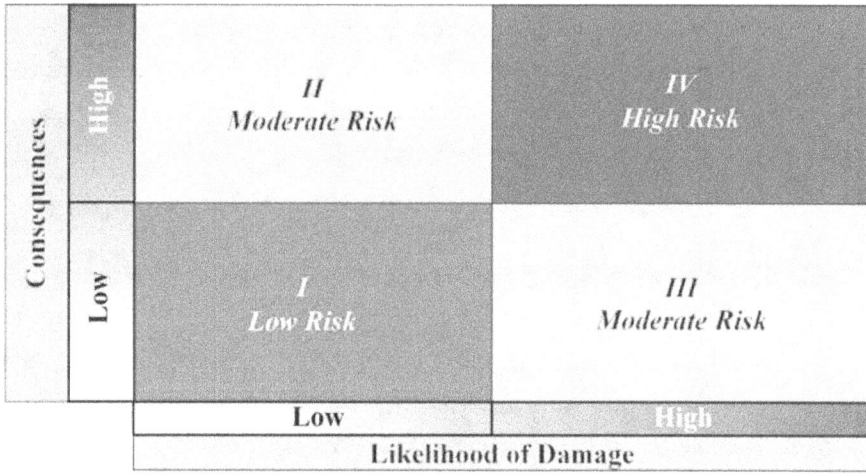

Figure 2-1   Seismic risk, expressed graphically as a function of likelihood of damage and consequences given the occurrence of the damage.

○ A national chain retail store in seismically active northern California; the building has been designed to perform well during severe earthquake ground motions and potential loss of use of one building out of hundreds would not disastrously affect the owner's business.

○ An abandoned warehouse in Texas; the seismic hazard is extremely low and the value to the owner is small.

Quadrant II, Moderate Risk: low likelihood of damage and high consequences; an example is:

○ A well-designed hospital in South Carolina; the probability of severe earthquake ground motions is low but the hospital has 100 critical care beds and an occupancy of 2,000.

Quadrant III, Moderate Risk: high likelihood of damage and low consequences; an example is:

○ A small storage facility for a national distributor located in a high seismic zone and designed to pre-1950 standards. The building is vulnerable to damage but the loss would likely be relatively unimportant to the owner.

Quadrant IV, High Risk: high likelihood of damage and high consequences; examples include:

○ A private day care center designed by an inexperienced engineer two miles from an active fault in a highly seismic region. Lack of knowledge of the hazards associated with near fault sites could result in injury to dozens of children.

○ A high-tech chip manufacturing plant in southern California, designed and built to the minimum requirements of the current code. The likelihood that a code-minimum building will experience extensive non-structural damage is high and business interruption could be devastating to the owner.

2.2 SEISMIC RISK MANAGEMENT: A HOLISTIC APPROACH FOR REDUCING EARTHQUAKE IMPACTS

Seismic risk management is simply the act of managing activities and decision making relating to building design, construction, and operations so as to reduce the impact of earthquakes.

Seismic Risk Management

Seismic risk management is the act of managing activities and decision making relating to building design, construction, and operations so as to reduce the impact of earthquakes.

One of the purposes of this document is to provide building design professionals with tools and strategies to help owners and managers make cost-effective seismic-risk management decisions. The document therefore describes and compares various strategies, including reducing the likelihood of earthquake damage and reducing consequences, or both. The document also provides information on estimating future costs resulting from earthquake damage and other impacts as well as the costs to improve performance in future earthquakes.

The likelihood of earthquake damage is a function of the seismic hazard at the site and the seismic vulnerability of the building. Seismic hazard is addressed in the context of site selection and evaluation of site-specific earthquake-related hazards, as discussed in Chapter 3. Seismic vulnerability is addressed in the context of performance-based design, a relatively new tool (discussed in Chapter 4) that engineers can use to adjust up or down the earthquake resisting capacity of a building, depending on the desired performance in future earthquakes.

Design variables and issues affecting seismic performance, along with guidance for calculating the cost of improving performance, are provided in Chapter 5. Chapter 5 also discusses the cost of improved seismic performance versus the cost of future earthquake damage and loss, including indirect costs resulting, for example, from time out of service. The means to quantify these costs are also discussed. The key to making wise investment decisions, as discussed in Chapter 5, can be found in a three-step process that consists of:

○ quantifying the amount and likelihood of losses that buildings may suffer in future earthquakes,

- estimating the expected reduction in future losses that can be achieved through various risk management programs, including performance-based design, and

- calculating the costs of implementing these programs, and comparing them to the estimated reduction in losses.

As with any other investment, the building owner weighs the expected return against the possible risk of not achieving that return. Equally important is the need to weigh the cost of the lost opportunity if the investment is not made. Examples abound in decision making involving such trade-offs, whether they relate to earthquake risk or other matters.

Specific seismic risk management strategies that focus on reducing the consequences or losses associated with earthquake damage are addressed later in this chapter, in terms of financial or business strategies and response planning strategies. Examples of these strategies include:

- diversifying operations so that all of an owner's operations are not concentrated in vulnerable buildings,

- obtaining insurance or other financial instruments to cover potential losses,

- establishing options to lease or buy replacement space after an event, or to immediately bring in contractors for repairs, and

- implementing pre-event planning and developing post-earthquake response and recovery programs to speed the process of business resumption.

It is often more effective, but typically more costly, to reduce seismic risk by reducing the likelihood that the earthquake damage will occur. By reducing seismic vulnerability, the uncertainties associated with estimating consequences of the expected damage and responding after a significant event are lessened. However, the initial costs of providing improved performance may never be recovered if an earthquake doesn't occur during the functional life of the facility. Reducing seismic risk by reducing the estimated consequences of a damaging earthquake often involves lower spending on an annual basis or incurring costs to repair or restore functionality once the event occurs. Large investments are not needed up front. In this case, while the likelihood of damage is not

CASE STUDY

Examples of Risk Management Strategies

1. Most businesses, whether commercial, industrial, or non-profit, know that reducing workplace injuries reduces expected costs in the future. Experience shows that capital spent today to install safety equipment and ergonomic furniture, and to conduct safety training for employees, can generate a positive return on investment by preventing future claims and reducing insurance premiums.

2. When deciding on a structural system for a new building, an initial extra 10% investment may result in less damage in future earthquakes. The benefit of not having to suffer as high a loss of capital, contents, and business interruption over the building's life can be compared to the investment cost at a given discount rate to determine the return and value of the investment.

reduced, the intent is to reduce seismic risk by enabling a quicker response and recovery.

The move to a performance-based design philosophy is a significant advance that can assist in seismic risk management, if it can be efficiently implemented into the building code development and design process.

2.3 EVALUATING SEISMIC RISK CONSEQUENCES AS A BASIS FOR DEVELOPING A RISK MANAGEMENT PLAN

The first step in developing a seismic risk management plan is to determine the nature and magnitude of the current risks. For a building or group of buildings, structural analysis procedures can relate potential damage to the intensity of shaking for a certain size earthquake. As the size of the earthquake increases, so does the total potential direct and indirect loss. Although the size of the loss increases with increasing magnitude, the likelihood of experiencing the loss decreases with increasing magnitude as the probability of earthquake occurrence also decreases with increasing magnitude.

Based on the likelihood of potential losses, one can determine the presumed capability to manage loss. Some owners and managers might rely on government assistance in combination with in-house resources to cover potential losses. The limit of these funds to pay for recovery costs would define current manageable loss. Losses in excess of this limit would be catastrophic and threatening to the business or institution. Figure 2-2 demonstrates this concept; the horizontal line defines the boundary between manageable and catastrophic loss. The intersection of the horizontal manageable loss line with the potential total loss curve defines the likelihood or risk of catastrophic loss. If this risk is too high, it can be reduced by increasing the capability to manage loss (moving the horizontal line up in Figure 2-2) and by reducing the potential loss curve with a higher performance objective for the building. Note that in Figure 2-2, the likelihood of the potential loss occurring is directly related to the probability of the earthquake, i.e., a smaller magnitude event corresponds to a high potential for occurrence and a larger magnitude event corresponds to a low potential for occurrence.

The capability to manage risks depends on the combination of several investment strategies on the part of facilities owners and managers. The first source of recovery funding is out-of-pocket expenses using in-house

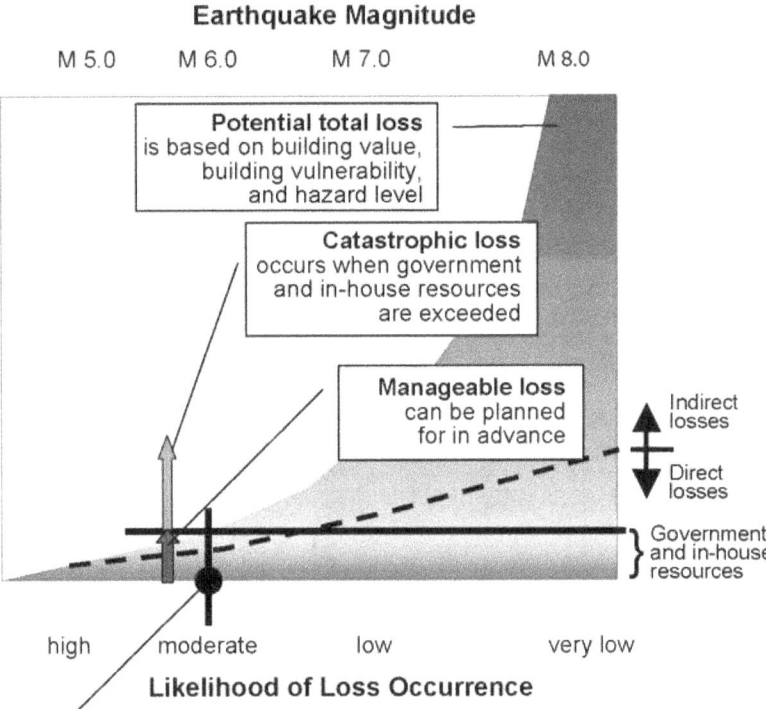

Earthquake Magnitude

M 5.0   M 6.0   M 7.0   M 8.0

Potential total loss
is based on building value,
building vulnerability,
and hazard level

Catastrophic loss
occurs when government
and in-house resources
are exceeded

Manageable loss
can be planned
for in advance

Indirect
losses

Direct
losses

Government
and in-house
resources

high   moderate   low   very low

Likelihood of Loss Occurrence

Risk that an earthquake will
cause catastrophic losses

Figure 2-2   Illustration of risk of experiencing catastrophic earthquake losses. Concept assumes building, or inventory of buildings, is located close to earthquake source region.

resources to cover losses. This may be supplemented by some sort of government disaster assistance. For example, Stanford University covered all of its costs resulting from the 1989 Loma Prieta earthquake with its own funds, supplemented by funds from the Federal Emergency Management Agency and the California State Office of Emergency Services. Small businesses may be able to obtain low interest recovery loans to increase their own resources.

Conventional insurance is a fairly common means of increasing manageable loss levels. This may be appropriate for smaller owners, whereas capacity might be a problem for a large institution such as a major university or hospital organization. Insurance rates fluctuate with the perceived market, and settlement delays can be quite costly in some cases. The capital markets may offer the flexibility to design financial instruments directly to suit an owner's specific needs using catastrophe bonds, which are effectively a combination of a loan and insurance. Conventional insurance and capital market investments can be used to increase the capability to manage loss. As discussed in Section 2.5, the optimal combination of these alternatives depends on insurance market conditions, interest rates, bonding capacity of the building owner, and other factors. Increasing the manageable loss level reduces the risk of catastrophic loss by elevating the horizontal loss limit line as illustrated in Figure 2-3.

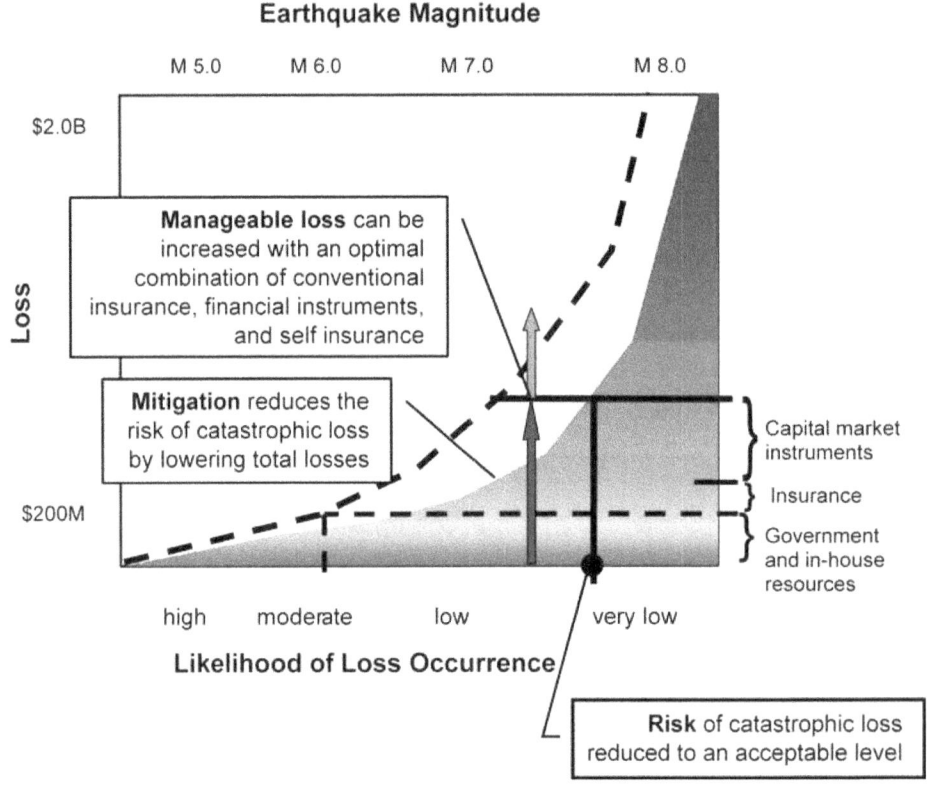

Figure 2-3   Illustration of reduction in risk of catastrophic earthquake losses.

Other strategies that can increase the level of manageable risk include the establishment of postearthquake response and recovery programs, which may reduce the amount of lost operations time through rapid engineering inspection and construction or repairs, or by obtaining alternate operating space quickly after an event. This is discussed further in Section 2.6.

Another important element of a risk management plan involves increasing the expected earthquake performance of the building, thereby lowering the potential loss curve. Mitigation reduces the risk of catastrophic loss by lowering the likelihood that the design earthquake would cause losses that exceed the manageable loss limit. The implementation of a mitigation strategy may include, as described in Section

2.4, designing new facilities to higher performance objectives, in order to limit losses over the building's life. This can apply to the replacement of outdated facilities or new facilities required as a result of company expansion needs.

The technical and financial parameters of a risk management plan all have associated uncertainties. Selecting the optimal combination of risk management strategies requires consideration of these uncertainties to assess the reliability of the decision making process. In addition, an integrated financial and technical model is necessary to test alternative strategies. The end result is a risk management plan that maximizes the return on investment to manage losses and reduce risk to an acceptable level over a fixed future time period. The flowchart shown in Figure 2-4 illustrates the various strategies that comprise a typical risk management plan and the options, or steps, for evaluating the strategies to select the optimal risk reduction solution. These three groups of strategies and related steps (outlined in Figure 2-4) are discussed in the next three sections.

2.4 FIRST COST OR DESIGN STRATEGIES

First cost or design risk reduction strategies are techniques that reduce the likelihood of damage to a structure. The term "first cost" is generally defined as an investment requiring a large capital outlay, whether or not it is truly spent near the start of a project. A capital investment of $10 million on a new building will most likely be amortized over some length of time, typically much longer than that actually required to construct the building. The owner is still responsible for the entire debt principal once the loan is secured, and often the debt goes "on the books" as a reduction in the amount of capital available for other investments.

First cost strategies reduce damage potential by either reducing the site hazards associated with a building or by increasing the expected performance of the building.

Reduce Site Hazards

An owner can reduce site hazards by reducing the intensity of earthquake shaking expected at the building site over the life of the structure, and by eliminating or reducing the potential for other seismic hazards, such as fault rupture, liquefaction, landslide, and inundation. Several techniques for accomplishing this are described below.

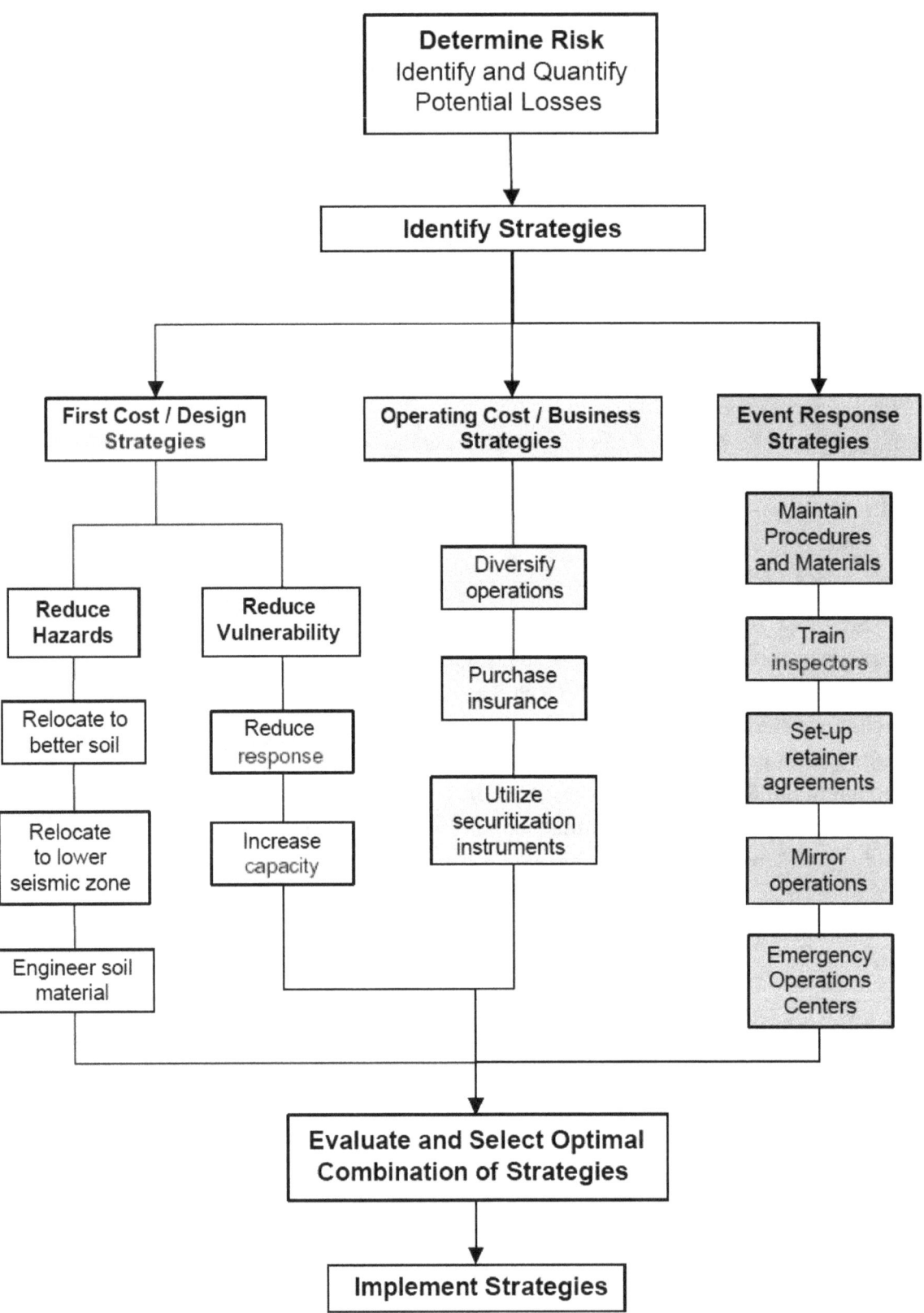

Figure 2-4   Flow chart for identifying, evaluating, and selecting risk-reduction strategies to develop a risk management plan.

○ *Locate the building in a region of lower seismicity,* where earthquakes occur less frequently or with typically smaller intensities. This option is generally the most effective strategy solely in terms of reducing the potential for earthquake damage to a facility, whether it be caused by ground shaking, fault rupture, liquefaction, landslide, or inundation. Locating a building in Dallas, Texas, for example, will almost guarantee that it will never be damaged by an earthquake. Of course, this option isn't possible for many building owners.

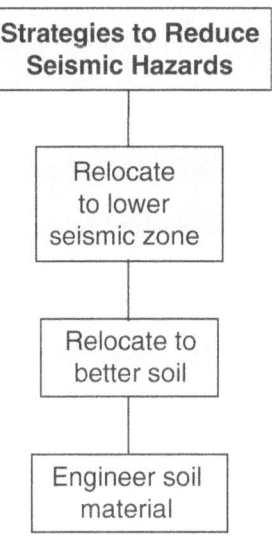

While certainly less desirable, and possibly quite costly from a market share and cost of operations standpoint, universities, manufacturing facilities, commercial offices and, to some degree, commercial retail owners, can use this strategy to manage their risks. Although it may not be practical for a university to build a new classroom facility across the country, locating some services off the main campus may be an option. For example, a university on the San Francisco peninsula located near the San Andreas Fault has considered siting a rare books depository approximately 75 miles south of campus, in an area of significantly less seismicity. It is also fairly common for high technology manufacturing plants to be located far from their headquarter locations, at sites with low seismicity such as Texas, Massachusetts, or Idaho. While it would be very rare for a retail establishment to make a siting decision based on seismic risk over the demographics of the market in a particular region, moving a store even a few miles in some cases can make a measurable difference in seismic hazard, e.g., moving a proposed building location from within a mile of a major fault to five miles away.

○ *Locate the building on a soil profile that reduces the hazard.* Local soil profiles can be highly variable, especially near water, on sloped surfaces, or close to faults. In an extreme case, siting on poor soils can lead to liquefaction, landsliding, or lateral spreading. Often, as was the case in the 1989 Loma Prieta earthquake near San Francisco, similar structures located less than a mile apart each performed in dramatically different ways because of differing soil conditions. Even when soil-related hazards are not present, the amplitude, duration, and frequency content of earthquake motions that have to travel through softer soils can be significantly different than those traveling through firm soils or rock.

An owner who is concerned about the effects of soil properties on risk should be encouraged to consult geotechnical and structural engineers to gauge the potential hazards associated with differing site conditions. These should be weighed against the costs, both direct and indirect, of locating the facility on soils that will result in better performance.

○ *Engineer the soil profile to increase building performance and reduce vulnerability.* If relocating to a region of lower seismicity or to an area with a better natural soil profile is not a cost effective option, the soil at the designated site can often be re-engineered to reduce the hazard. On a liquefiable site, for instance, the soil can be grouted or otherwise treated to reduce the likelihood of liquefaction occurring. Soft soils can be excavated and replaced, or combined with foreign materials to make them stiffer. The building foundation itself can be modified to account for the potential effects of the soil, reducing the building's susceptibility to damage even if liquefaction or limited landsliding does occur.

The owner should weigh the additional costs of modifying the soil profile or the building foundation (which may be quite significant in certain cases) with the expected reduction in damage and loss.

Improve Building Performance

An owner can reduce vulnerability by increasing the performance of the building, thereby reducing the damage expected in earthquakes. There are two methods by which this is typically accomplished:

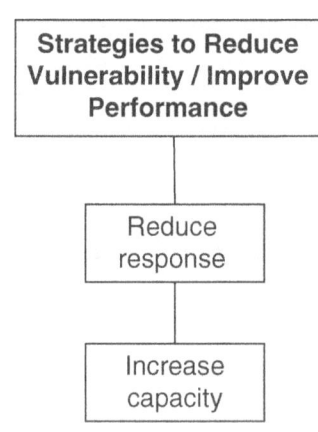

○ *Reduce the response of the building to earthquake shaking.* An earthquake generates inertial forces in a building that are a function of the structure's mass, stiffness, and damping, and of the acceleration and frequency of the earthquake motion. The parameters associated with the earthquake can only be altered by reducing hazards, as described above. While the actual mass of the building (the weight of the structure, contents, and people) typically cannot be significantly altered, the effective mass can be changed by providing special devices, such as passive or active mass dampers, that can effectively reduce the overall mass that is accelerated by the earthquake. Stiffness can be altered by modifying the structural system (e.g., concrete shear wall, steel moment frame) or by using braces and seismic dampers. The building's fundamental period, which is an important parameter in determining building response, can be significantly increased (and resulting forces reduced) by providing seismic isolating devices at the building foundation.

Engineers familiar with the use of these response-modifying devices can relatively easily quantify both the costs and benefits of employing them in buildings. When these types of products were new to the building industry, they were generally expensive. Today, with competition in the marketplace, they are much more common and costs have dropped dramatically.

○ *Increase the capacity of the building to resist earthquake forces.* The most traditional method for decreasing vulnerability of buildings is to make them "stronger." By increasing the forces that a building can resist, such as by providing larger structural elements or increasing the amount of bracing for nonstructural systems, less damage would be expected. This strategy can be costly and, in some cases, may not be the most efficient means of increasing performance. Another option is to increase the ductility of the structural or nonstructural systems, improving their ability to absorb the energy of the earthquake without permanent damage.

Increasing the capacity of the building may be the most difficult strategy to quantify reliably because of the inherent complexity of most structural and nonstructural systems. However, the range of possible solutions (and costs) for increasing capacity is relatively large, thus this strategy is the often employed because it allows the engineer to fine-tune a design approach to meet an owner's budget and risk management criteria.

2.5 OPERATING COST OR BUSINESS STRATEGIES

Operating cost or business risk reduction strategies are techniques that primarily enhance the capacity to manage losses, by effectively reducing the consequences of damage. The term "operating cost" is generally defined as an investment made on an annual or other regular basis.

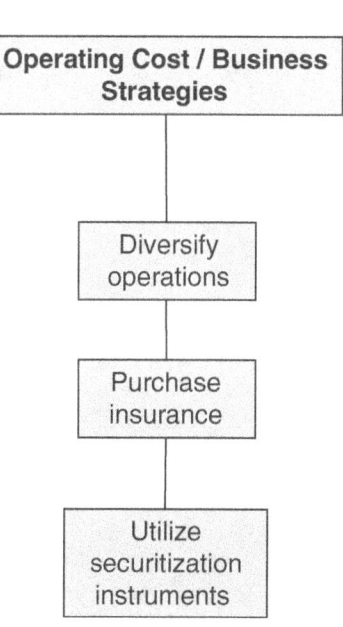

Diversify Operations

An owner with geographically-dispersed buildings, or with an inventory consisting of buildings of various ages and seismic performance characteristics, can reduce overall earthquake risk by moving certain operations to buildings located in regions of lower seismic hazard or to buildings of higher seismic performance. This strategy can be fine-tuned when different operations carry different earthquake risks in terms of business disruption, loss of contents, or other impacts. For example, high resource operations, such as manufacturing or administration, can be relocated to new, higher performing buildings, while archival storage can be moved to older, more vulnerable ones. This can be done incrementally as new buildings are brought on line or over a defined timeframe so as to minimize the impact on operations.

Consider the following example of two manufacturing businesses. One runs 100% of its production from a single building in San Francisco. The other runs 50% each from one building in San Francisco and from one building in Seattle. There is a one percent annual chance in each city of an earthquake large enough to cause complete loss to the buildings.

Company A
100% operations in San Francisco
1% annual risk of complete loss to San Francisco building
Overall result: a 1% annual risk of complete business loss

Company B
50% operations in San Francisco
50% operations in Seattle
1% annual risk of complete loss to San Francisco building
1% annual risk of complete loss to Seattle building

Overall result:

- *a 1% annual risk of 50% business loss due to San Francisco event*
- *a 1% annual risk of 50% business loss due to Seattle event*
- *a 2% annual risk of 50% business loss*
- *a 0.01% annual risk of complete business loss (1% × 1%)*

As the number of sites grows, the risk of suffering a catastrophic loss to the business drops exponentially, even though the risk of suffering some loss grows. This assumes, of course, that the sites are independent. Having two similar buildings in San Francisco, within a mile of each other might not decrease the risk as substantially since a single earthquake could affect both. This methodology is used by insurance companies regularly to spread out their risk and reduce the potential for a single disaster causing more claims than they can settle.

Obtain Higher Levels of Insurance

An organization should ensure that it has a sufficient amount, and type, of insurance coverage to adequately protect against losses. This determination is typically made by an owner's risk manager (or the insurance broker, acting on behalf of the owner). The risk manager must assess the cost of insurance relative to the potential costs of accepting the risk without insurance coverage. In most cases, investments in risk reduction (e.g., improving building performance or relocating to a lower risk area) will also result in insurance premium reductions. The risk manager must balance these different options by assessing the life cycle costs and benefits of each option. In order for the overall risk management plan to be effective, it is important for the organization's risk manager to become an integral part of the management team making facility decisions, and that communications with the facility manager and the design team be open and complete.

Using Securitization Instruments

Conventional insurance is typically best suited for incidents that occur regularly, although possibly infrequently, such as fire and worker injury. Conventional insurance is also appropriate for losses that are easily quantifiable, such as losses to capital and inventory. For very rare or catastrophic losses, however, obtaining insurance coverage can often be cumbersome or costly. It becomes difficult to price insurance when the rates of incidents are uncertain and when coverage for indirect losses from business interruption or loss of market share is needed.

A newer instrument, commonly called a Catastrophe Bond (Cat Bond) has recently garnered some attention. Sold on the open market, the proceeds of these bonds (typically in the range of $10 million to $100 million) are held in a financially secure trust. If an earthquake occurs within the period of the bond, and if the earthquake meets certain criteria in terms of size, location, or loss, some or all of the bond's principal or interest is forfeited to the seller to assist recovery. While generally limited to reinsurance companies, a small number of large, private corporations have started offering Cat Bonds. When insurance rates are low, Cat Bonds are generally less attractive. However, when insurance rates are high, as is common after a disaster, bonds become more economical.

Large investment banks are generally the best source to help an owner explore securitization options. The owner should have adequate understanding of his or her expected losses in different events, however, so that the amount of the bond and the interest payments can be as small as possible and yet still cover potentially catastrophic damage. Design team members well versed in performance-based design, risk assessment, and loss estimation, can be a valuable resource to owners in this effort.

2.6 EVENT RESPONSE STRATEGIES

The goal of event response risk reduction strategies is to manage potential losses through quick recovery and response to damage caused by earthquakes. Similar to operating cost or business strategies, event response focuses on reducing the consequence of damage and loss, rather than reducing the likelihood of damage and loss occurring. Relative to first cost or design strategies, event response typically requires much lower initial costs, as well as lower annual operational costs. While event response does not typically reduce capital losses or the amount of repair that may be needed, it can speed up the process of recovery through effective pre-event planning.

Emergency Response Procedures and Materials

The simplest form of an event response strategy can consist of maintaining procedures, equipment, and materials on-site for aiding the evacuation of building tenants. Most companies and institutions have at least a basic emergency kit and response procedures for evacuating people from a potentially hazardous building after an earthquake or during a fire. This level of planning can be implemented at a minimal cost. It may aid in the evacuation of the building and the treatment of injuries, but will not reduce capital or business interruption losses.

Pre-Event Disaster Training and Inspection Services

A second strategy is to develop and provide formal disaster training for employees and building personnel. Many large companies have instituted basic emergency response training for their employees, which includes basic safety and medical training. It may also include primer level education on how buildings respond in earthquakes and what telltale signs of building damage might indicate potential safety or operational hazards. These programs are generally not technically sophisticated, and are not intended to be a substitute for professional emergency response or engineering personnel. They can, however, reduce lost time in the event that damage is minimal and if building occupants are efficiently organized in the recovery effort.

A key component of event response is the ability to adequately identify what building damage means in terms of occupant safety and building functionality. To be done reliably, this requires the use of professional engineers and architects who have taken comprehensive training in the evaluation of earthquake damaged structures. Building owners and tenants can not reasonably be expected to accept the liability of deciding

---

Event Response Strategies

- Maintain Procedures and Materials
- Train inspectors
- Set-up retainer agreements
- Mirror operations
- Emergency Operations Centers

whether a building is safe to occupy. Thus, post-event engineering inspections are an important tool in an overall event response strategy. Large organizations will typically keep architects or structural, mechanical, and electrical engineers on retainer to quickly respond after a major event. They will be authorized to make safety inspections of owner's facilities using guidelines typically established by the local jurisdiction, such as the ATC-20 post-earthquake building safety evaluation procedures (ATC, 1989, 1995, 1996). They will then make recommendations to the building owner and put up signs noting whether the building is safe to enter, unsafe, or has restricted access in some fashion.

For owners or tenants of several buildings, this strategy should include the entire network of buildings that could be affected by an earthquake. Inspectors will be used most efficiently if they are sent to the buildings that are most severely damaged. For critical facilities, such as hospitals, it is advantageous to predict in advance the expected safety inspection postings—INSPECTED (green placard), RESTRICTED USE (yellow placard), and UNSAFE (red placard)—for all buildings on site. Procedures to be followed in developing and using such posting predictions are provided in the ATC-51-1 report, *Recommended U.S.-Italy Collaborative Procedures for Earthquake Emergency Response Planning for Hospitals in Italy* (ATC, 2002)

Pre-Event Disaster Training and Inspection

1. ATC-20, *Procedures for Postearthquake Safety Evaluation of Buildings* (ATC, 1989),
2. ATC-20-1, *Field Manual: Postearthquake Safety Evaluation of Buildings* (ATC, 1989),
3. ATC-20-2, *Addendum to the ATC-20 Postearthquake Building Safety Evaluation Procedures* (ATC, 1995).
4. ATC-20-3, *Case Studies in Rapid Postearthquake Safety Evaluation of Buildings*, (ATC, 1996)
5. ATC-51-1, *Recommended U.S.-Italy Collaborative Procedures for Earthquake Emergency Response Planning for Hospitals in Italy* (ATC, 2002)

On-Retainer Temporary Space and Repair Contractors

A relatively new strategy being employed by some businesses is to obtain disaster lease or repair options for their facilities. Organizations may execute agreements with local contractors and property owners to provide them with a choice of available temporary space and for the labor necessary to repair damaged facilities. If an organization's buildings are damaged to the point that they are not functional and need significant repair, the owner would have the first right of refusal on any available space that a landlord has, at an agreed-upon set of pre-event prices, and on the use of personnel from local contractors. The organization, in return, provides a yearly retainer for the service.

Contingency planning companies offer building owners a service referred to as a "hot site." This is typically a fully equipped and functional facility (usually office space) that can be occupied fairly rapidly, whenever the owner's facility becomes unusable as the result of a natural or man-made disaster. Such a service can significantly reduce or eliminate the costs of business disruption resulting from earthquake damage. The service does not usually include the repair of the owner's damaged facility.

Facility and Data Mirroring

A formal business occupancy resumption program may include developing procedures by which critical information and the ability to conduct business are backed up or duplicated at alternate sites. This can range from electronically backing up and storing computer data at an off-site location to supplying company-wide transportation assistance to and from the employees' place of business, to having a plan to swiftly relocate operations to other offices or locations. Depending on the nature of the business, one or more of these options may be applicable.

Emergency Operations Centers

An emergency operations center is typically a hardened facility in which managers can conduct the emergency response and recovery effort. This facility will be constructed or located so that it can be operational after a major event. It should house emergency communications equipment, information on buildings and their contents, and have access to maps and information detailing the extent of damage both to the owner's facilities and the surrounding areas. The emergency operations center should be stocked so as to remain operational for at least 72 hours. For owners with large inventories of buildings or where a complex network of inspection and recovery is needed, the creation of an emergency operations center can be an effective strategy to ensure that building safety is rapidly evaluated and business resumption can occur as soon as it is safe and possible to do so.

2.7 CHOOSING AN OPTIMAL COMBINATION OF RISK REDUCTION STRATEGIES

Choosing an optimal combination of risk reduction strategies among those described above involves weighing the costs and potential savings associated with each option. The goal is to determine which combination results in the best return on investment. The basic steps that should be taken to identify an optimal combination of risk reduction strategies include the following.

Steps to Identify an Optimal Combination of Risk Reduction Strategies

1. Identify potential losses
2. Quantify losses
3. Identify risk reduction strategies
4. Select and implement strategies

1. *Identify potential losses.* These losses include, as described earlier, capital, contents, casualties, business interruption, and market share. A qualitative description of the type of damage that could be suffered is an appropriate starting point. This should include the use of engineering evaluation and performance-based design (see Chapter 4). For a range of earthquake scenarios or probabilities, a description of the type of capital and contents damage, estimates of casualties, and estimates of the duration of business interruption can be made by evaluating the structural and nonstructural behavior of the building under the given shaking intensity (See Section 5.7 for additional discussion). The design team and facilities staff should work as a team to identify, as examples, how long building operations will be shut down if shear walls in a building are cracked to the extent that structural repair is necessary, or what the average continuous occupancy of a classroom is over the course of a month.

2. *Quantify losses.* After the losses are qualitatively identified, they need to be translated into a common quantitative measure. The total value of both anticipated direct and indirect losses should be determined. In some cases, indirect losses will need to be converted into a direct cost equivalent through a value-based conversion; e.g., total manufacturing days lost, or hospital bed days lost.

3. *Identify risk reduction strategies.* Once losses have been quantified, the design team and owner's representatives should explore methods for reducing these losses, using the strategies described above; i.e., first cost or design strategies, operating cost or business strategies, or event response strategies. The team should identify as many options within each method as practical, estimating the cost to implement each and the expected savings in terms of reduced losses.

For first cost or design strategies, various performance objectives should be considered for the new building. For example, the baseline scheme would be a building that meets the minimum provisions of the applicable building code. A higher performance objective might be one in which the building is not necessarily functional after the design event, but in which operations can be restored within a relatively short period of time. A still higher performance objective may be a building that is designed to remain fully functional in the design event. For each higher performance option, a conceptual level of structural and nonstructural design should be performed to determine an approximate cost difference above the baseline option. It is generally sufficient to make rough order approximations of costs.

For each design strategy, the performance of the building is then translated into an expected loss (both direct and indirect) in a range of earthquake scenarios with different probabilities of occurrence. To facilitate comparisons, these should be translated into an expected annual loss and converted to a present value at an assumed discount rate.

For operating cost or business strategies, different options such as insurance, securitization, and lease/repair options should be explored. The annual premiums to obtain specified amounts of insurance can be calculated with the help of the owner's insurance brokers. It is important to understand whether the insurance will also cover business interruption losses, or only capital and contents. For larger owners, catastrophe linked securities can also be considered. Typically, the coverage provided by such securities ranges from $10 million to $100 million, so smaller entities would not find these products appropriate. Lease/repair options can be developed with contractors and landlords in the nearby vicinity of the building, typically for an annual retainer fee. For all of these options, annual costs should be converted to a present value in order to facilitate a comparison with the other strategies.

For event response strategies, the design team should develop an outline of some specific options, such as a post-earthquake inspection program or the establishment of an emergency operations center. Annual or initial costs can be relatively easily estimated from this, as they will typically be small relative to the other strategies. The main benefit of this strategy will be to reduce down time as a result of better pre-event planning. The resulting savings should be approximated in terms of the daily cost of lost operations, multiplied by the expected reduction in lost time. These costs and savings should also be annualized and converted to a present value for comparison purposes.

4. *Select and implement strategies.* Once the various options within each strategy group have been evaluated and quantified, the team should consider permutations of each to determine the ones with the overall greatest benefit-to-cost ratios. Other factors may make certain strategies less desirable (such as the difficulty of a local school commission passing a bond measure for a seismic improvement capital outlay). Where appropriate, each option should be given a weight-

ing factor to express its desirability apart from purely economic factors. Once the optimal combination of strategies is selected, the owner and design team should develop a plan to implement them as part of the overall design process.

2.8 EXAMPLE IMPLEMENTATION OF A RISK MANAGEMENT PROGRAM

CASE STUDY

The following example illustrates how a hypothetical company might develop a risk management program.

Description of company

A Hayward, California company manufactures computer memory boards for large computer makers. The company's annual revenues are $100 million. All manufacturing is done in a single building with administration in a separate facility located in the same office park about three miles from the Hayward Fault. The company is planning to expand in the next three years, to double its annual revenues. It will buy a second manufacturing facility and construct a new office building.

Establishment of risk tolerance

Because the company currently has only four major clients it has determined that the risk of losing a substantial amount of revenue due to being dropped by a client is intolerable. It has decided to permit no more than 5% annual chance that lost client revenue exceed 10% of its revenue. Being near the Hayward Fault, the company concludes that it should not arbitrarily tolerate a large earthquake risk.

Current seismic risks

The company has conducted an engineering analysis of its current buildings and calculated the seismic risk. In a Hayward Fault event with a 5% annual chance of exceedence, capital and contents losses could reach $10 million. Loss of operations could cost another $25 million. The total direct losses of $35 million exceeds 10% of the company's revenue.

Risk Management Strategies

First cost / design strategies – The company can either buy new properties in the same office park or across the San Francisco Bay. By grouping all the buildings in the same location, the seismic hazard remains unchanged. However, the consequence of an earthquake on the Hayward Fault is increased because all of the buildings are likely to be affected. By separating the buildings geographically, the vulnerability of all four buildings being damaged by an earthquake in a year becomes negligible, although the chance that at least two of them will be affected by any event goes up. Engineers calculate that diversifying the regional location of operations reduces the overall risk most effectively. This results in reduced capital, contents and business interruption losses associated with a 5% annual chance of exceedence to $31 million, or 15.5% of the now doubled $200 million annual revenues.

Operating cost / business strategies – The company considers two operating cost options to further reduce the risk to its target tolerance. It can either obtain insurance to cover the remaining losses not managed by the improved first cost strategies, or it can implement a program of incremental retrofit of the two manufacturing buildings, to reduce their vulnerability. The company decides that the most cost effective option is to obtain insurance to cover (after deductibles) $5,000,000 of remaining potential losses above its tolerance. This represents 2.5% of the company's annual revenues

Event response strategies – The company decides to establish a post-earthquake response program, whereby it contracts with local engineers and contractors to provide immediate post-event inspections and repair design. The annual cost of developing and maintaining the program is $40,000 per year for the four buildings. The company conservatively estimates that in a moderate-to-large earthquake it could save at least a week-and-a-half of lost operations by having an engineer on board immediately. This equates to $6 million, or 3% of the annual revenue.

The three strategies together result in the following risk management program that meets the company's tolerance.

Vulnerability assumes hazard with 5% annual chance of exceedence:

First cost strategy: Total losses = $31 million (15.5% of revenues)

Operating cost strategy: Reduces total losses by $5 million, to $24 million (11% of revenues)

Event response strategy: Reduces business interruption and total losses by $6 million, to $20 million (10% of revenues)

2.9 SEISMIC RISK MANAGEMENT ADVOCACY

Corporate cultures, especially those related to perception and tolerance of risk, are difficult to change. As a result, encouraging a corporate or institutional mindset to place a higher emphasis on seismic risk management will usually require one or more "champions" who can work from both within and outside the organization.

From within an organization, the organization's risk manager or a top level staff member from the facilities department is the likely most appropriate in-house champion, as these individuals are likely to have both a broad understanding of the company's corporate and business goals, and detailed knowledge of the design and construction process. The in-house champion will be expected to introduce seismic risk management standards, establish design priorities, quantify the consequences of losses, develop ongoing risk reduction processes, and keep the facilities department staff aware of their activities and findings. This champion must also be able to persuade upper level management of the need for such changes in policies and procedures. This will often require that the in-house champion "speak" in two languages – one technical and the other financial.

The design team should also act to champion the seismic risk management cause at early stages of discussion regarding a new building. In current practice, most design teams are organized under the direction of the project architect, who typically has a direct relationship to the building owner. Therefore, the external champion may be someone from within the project architect's office.

It is important that both internal and external champions believe that seismic risk management truly adds value to their services and to the overall design process. Additionally, the external champion should establish a relationship with the internal champion, in order to leverage and support each others efforts, and to further the risk management process.

IDENTIFYING AND ASSESSING EARTHQUAKE-RELATED HAZARDS 3

3.1 INTRODUCTION

This chapter provides an overview of earthquake-related hazards affecting buildings as well as guidance on how to consider these hazards in the site selection process for new buildings. While seismic shaking is potentially the greatest threat, the collateral seismic hazards of fault rupture, liquefaction, soil differential compaction, landsliding, and flooding (inundation) could also potentially occur at a site. In addition, there are other hazards associated with the built environment that may affect building performance in the earthquake aftermath. These include: (1) hazards arising from external conditions to the site, such as vulnerable lifelines (transportation, communication, and utility networks) and hazardous adjacent structures, including buildings close enough to pound against the building that is to be constructed at the site; (2) storage and distribution of hazardous materials, and (3) postearthquake fires.

Section 3.2 discusses seismic shaking hazards, including the current technical and code approaches for quantifying the shaking hazard. Section 3.3 identifies and discusses the collateral seismic hazards that should be considered in selecting an appropriate site for a new building (fault rupture, liquefaction, soil differential compaction, landsliding, and flooding). The other collateral hazards that could affect site selection decisions (vulnerable lifelines, hazardous adjacent structures, storage and distribution of hazardous materials, and postearthquake fires) are discussed in Section 3.4. Specific guidance on actions to be taken to assess earthquake-related hazards during the site selection process, including a checklist for site analysis, are provided in Section 3.5. Resources for further reading are provided in Section 3.6. All sections are written in technical terminology appropriate for design professionals to aid in communicating with building owners and managers.

3.2 EARTHQUAKE GROUND SHAKING HAZARD

The effects of ground shaking on building response are well known and extensively documented. Severe ground shaking can significantly damage buildings designed in accordance with seismic codes (Figures 3-1 and 3-2) and cause the collapse of buildings with inadequate seismic resistance (Figures 3-3 and 3-4).

Figure 3-1   Six-story concrete-moment-frame medical building that was severely damaged by the magnitude-6.8 Northridge, California, earthquake of January 17, 1994. The building was subsequently demolished without removing contents. (photo courtesy of the Earthquake Engineering Research Center, University of California at Berkeley)

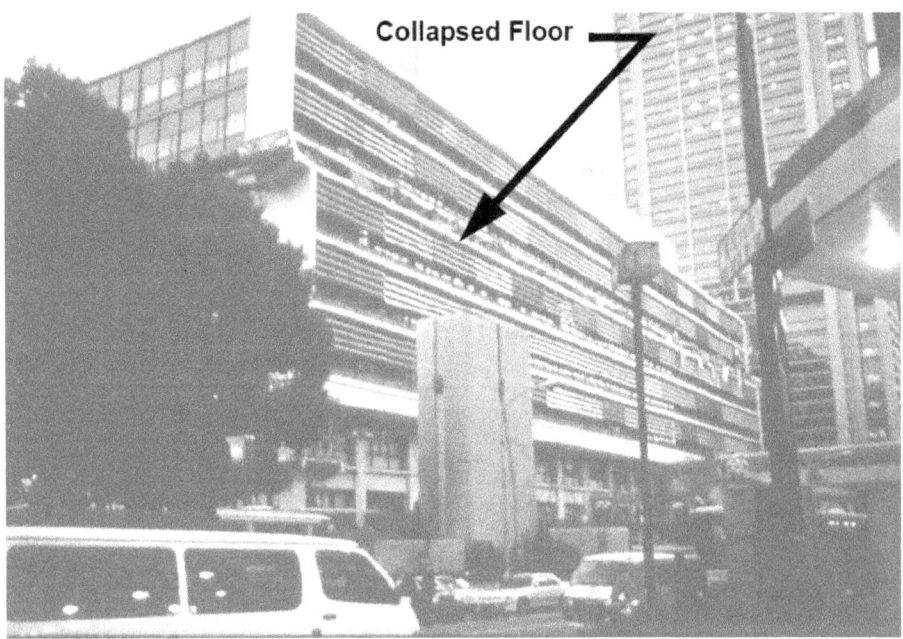

Figure 3-2   Eight-story reinforced-concrete-frame office building in Kobe, Japan that partially collapsed during the magnitude-7.8 earthquake of January 17, 1995. Note that the sixth floor is missing, due to collapsed columns at that level. Seismic codes in Japan are essentially equivalent to those in the United States. (photo courtesy of C. Rojahn)

Figure 3-3   Older five-story reinforced concrete frame building in Managua, Nicaragua, that had inadequate seismic resistance and collapsed during the magnitude-6.2 earthquake of December 23, 1972. (photo courtesy of C. Rojahn)

Figure 3-4   Preseismic-code ten-story reinforced-concrete-frame building in Bucharest, Romania, that partially collapsed during the 1977 magnitude-7.2 earthquake approximately 65 miles north of Bucharest. (photo courtesy of C. Rojahn)

Quantifying the Earthquake Ground Shaking Hazard

Seismic shaking is typically quantified using a parameter of motion, such as acceleration, velocity, or displacement. In current seismic codes, seismic design forces are defined in terms that relate to acceleration in the horizontal direction. A typical acceleration time-history of strong ground shaking is shown in Figure 3-5.

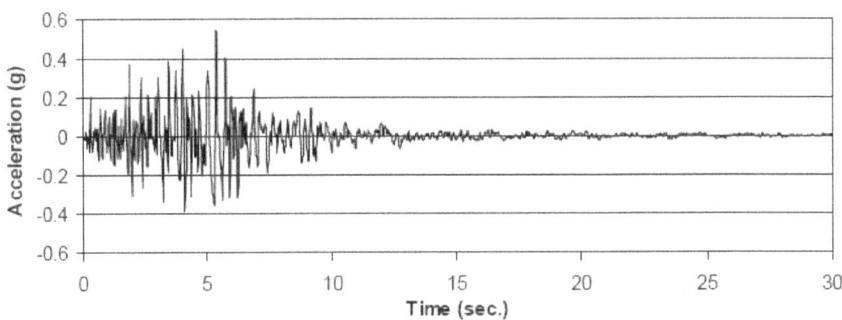

Figure 3-5   Typical acceleration time history of strong ground shaking.

The earthquake ground shaking hazard for a given region or site can be determined in two ways: deterministically or probabilistically. A deterministic hazard assessment estimates the level of shaking, including the uncertainty in the assessment, at the building site for a selected or scenario earthquake. Typically, that earthquake is selected as the maximum-magnitude earthquake considered to be capable of occurring on an identified active earthquake fault; this maximum-magnitude earthquake is termed a characteristic earthquake. A deterministic analysis is often made when there is a well-defined active fault for which there is a sufficiently high probability of a characteristic earthquake occurring during the life of the building. The known past occurrence of such an earthquake, or geologic evidence of the periodic occurrence of such earthquakes in the past, are often considered to be indicative of a high probability for a future repeat occurrence of the event.

The earthquake ground shaking hazard for a given site can be determined in two ways: deterministically or probabilistically. A deterministic hazard assessment estimates the level of shaking at the building site for a selected or scenario earthquake. Probabilistic hazard assessment expresses the level of ground shaking at the site with a specific probability of being exceeded in a selected time frame (normally 50 years)

Probabilistic hazard assessment expresses the level of ground shaking with a specific, low probability of being exceeded in a selected time period, for example 10% probability of being exceeded in 50 years, or 2% probability of being exceeded in 50 years, where 50 years is commonly chosen as the building design life. The seismic loading criteria in current U.S. building codes define design force levels based on ground motions specified in probabilistic seismic hazard maps. Such

IDENTIFYING AND ASSESSING EARTHQUAKE- RELATED HAZARDS

maps include those showing expected peak ground acceleration and those showing expected peak spectral acceleration response at different building periods of vibration. Figure 3-6, which was prepared by the U.S. Geological Survey National Seismic Hazard Mapping Project, illustrates a probabilistic seismic hazard map showing the regional variation of ground shaking hazard in the contiguous United States.

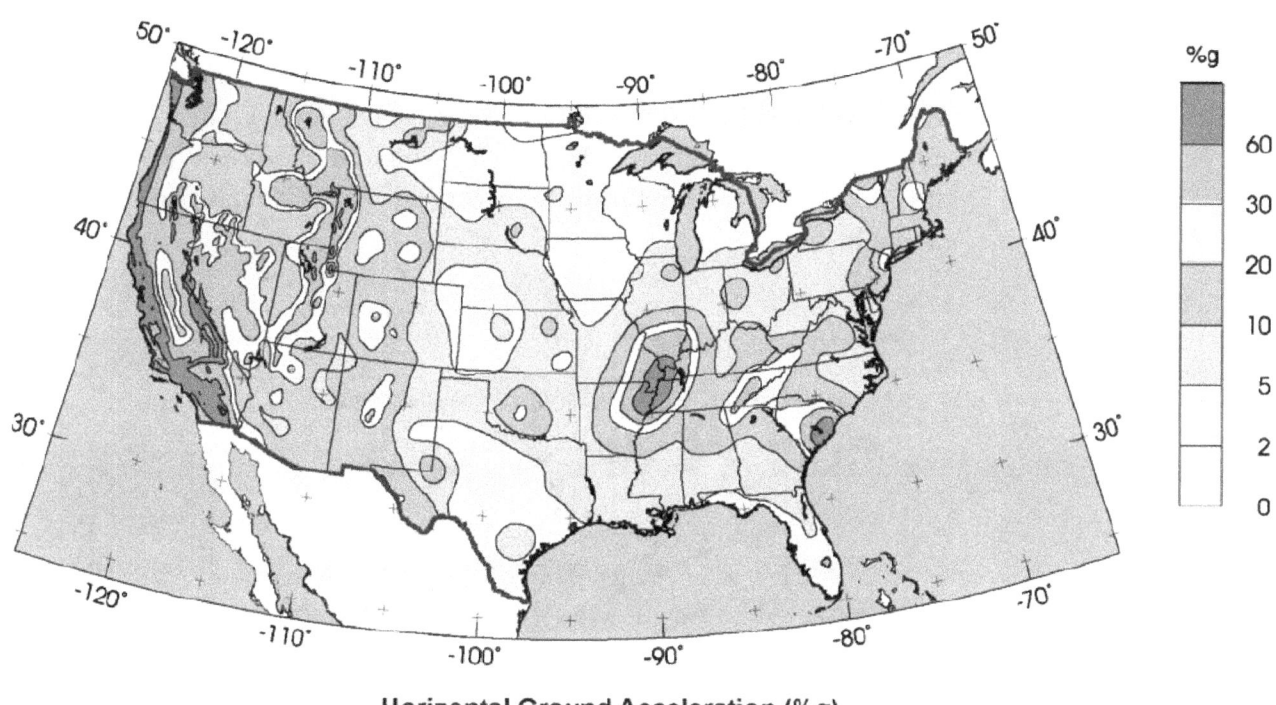

**Horizontal Ground Acceleration (%g)
with 2% Probability of Exceedance in 50 Years**

Figure 3-6   Probabilistic seismic hazard maps showing ground shaking hazard zones in the contiguous United States. (from USGS National Seismic Hazard Mapping Project website: geohazards.cr.usgs.gov).

This map indicates that, although the level of earthquake activity is high in California, most parts of the United States are also exposed to a significant earthquake ground shaking hazard. In fact, large historic earthquakes in the United States have occurred outside California, in Missouri, Arkansas, South Carolina, Nevada, Idaho, Montana, Washington, Alaska, and Hawaii. Furthermore, current geologic studies have shown increasing evidence for large earthquake potential in areas that are popularly believed to be relatively quiet. Examples include the now-recognized subduction zones in Oregon and Washington, the Wasatch fault zone in Utah, and the Wabash Valley seismic zone in Illinois and Indiana.

RISK CONSIDERATION

Although the level of earthquake activity is high in California, most parts of the United States are also exposed to a significant earthquake ground shaking hazard.

Probabilistic estimates of ground shaking at a given site can also be determined from a probabilistic ground shaking analysis for the site (often termed a "probabilistic seismic hazard analysis" or PSHA), whereby a geotechnical engineer determines and integrates contributions to the probability of exceedance of a ground motion level from all earthquake faults and magnitudes that could produce potentially damaging ground shaking at the site. Figure 3-7 illustrates relationships, termed "hazard curves," which indicate the level of peak ground acceleration and annual frequency of exceedance for specified locations in seven major cities in the United States (which have been obtained from the USGS National Seismic Hazard Mapping Project). From relationships such as those shown in Figure 3-7, ground motions can be readily obtained for any selected probability of exceedance and building design life.

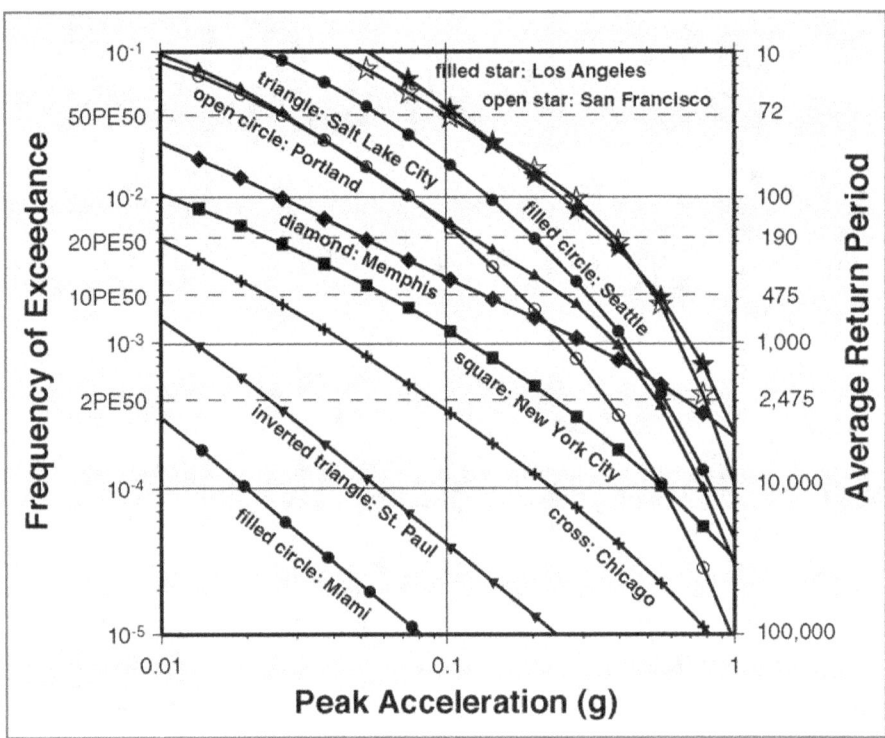

Figure 3-7   Hazard curves for selected U.S. cities.

For applications in performance-based design (see Chapter 4), both a probabilistic approach and a deterministic approach for the ground shaking hazard assessment may be used. Using a probabilistic approach, the seismic hazard can be integrated with the building resistance characteristics to estimate expected damage or loss, or the probability of exceeding some level of damage or loss, during a time period of significance such as the anticipated building life or the period during which the building will have a particular use. Using a deterministic approach, the expected damage or loss, or the probability of exceeding either a specified damage level or a specified loss, may be assessed for an earthquake considered to be sufficiently likely that satisfactory building performance during the earthquake is desired.

Determining Design Ground Motions for a Specific Site

Design ground motion for a given site can be obtained from national ground motion maps, such as the map shown in Figure 3-6, which defines ground shaking for a reference (standard) rock condition. When using the national ground motion maps (e.g., Figure 3-6) to define design ground motions for a given site, published soil factors are used to adjust the mapped values to reflect the soil conditions at the site. National ground motion maps include purely probabilistic hazard representations (peak acceleration response with a 2% probability of exceedance in 50 years, or 10% probability of exceedance in 50 years), as developed by the U.S. Geological Survey (Frankel et al., 1996; Frankel et al., 2000; Frankel and Leyendecker, 2000). Maps of modified levels of these hazards incorporate deterministic bounds on ground motions near highly active faults. Maps containing deterministic bounds, which are termed Maximum Considered Earthquake (MCE) maps, are found in the 2000 *NEHRP Recommended Provisions for Seismic Regulations for New Buildings and Other Structures* (FEMA 368 report) and its companion Commentary (FEMA 369 report), or in the 2000 *International Building Code* (*IBC*). Site factors to adjust the level of ground shaking from the reference rock condition to various softer soil conditions are also contained in the FEMA 368 *NEHRP Recommended Provisions for Seismic Regulations for New Buildings and Other Structures* (BSSC, 2001) and the *IBC* (ICC, 2000).

Site-specific studies can also be done to supplement or bypass the national ground motion maps. Such studies are most often undertaken for sites having soft soil conditions not covered by site factors published in the EMA 368 *NEHRP Recommended Provisions for Seismic Regulations for New Buildings and Other Structures* or the *IBC*, for sites close to earthquake faults, and for buildings considered to be of sufficient importance to warrant additional focus on regional and area-specific factors affecting ground shaking. Site-specific studies offer the potential for a more detailed analysis of the uncertainty in the seismic ground shaking hazard, as discussed below. If site-specific studies are conducted, they should be comprehensive and be subjected to detailed peer review.

Uncertainty in Hazard Assessment

Whether the seismic shaking hazard is estimated probabilistically or deterministically, there is always uncertainty in the hazard assessment and in the assessment of building performance. To provide a robust assessment of hazard, it is important to incorporate the uncertainty in aspects such as:

- ○ magnitude of the largest (i.e., characteristic) earthquake that can occur on an earthquake fault;
- ○ recurrence rates of earthquakes of different magnitudes on a fault;
- ○ the most applicable ground-motion-estimation relationship for a particular site, given the available models published in the technical literature; and
- ○ site response effects.

Each of these examples of uncertainty will have a different impact on a seismic ground shaking hazard assessment, and studies to assess the sensitivity of the hazard uncertainty on building performance are often conducted by multidisciplinary teams containing both seismologists and engineers.

3.3 COLLATERAL SEISMIC HAZARDS

RISK CONSIDERATION

Most current seismic design codes are not intended to prevent damage due to surface fault rupture; liquefaction, landslides, ground subsidence, or inundation.

In addition to strong ground shaking, there are other (collateral) seismic hazards – surface fault rupture, soil liquefaction, soil differential compaction, landsliding, and flooding (inundation) – that are potentially so severe that they could impact development costs to such a degree as to cause the site to be rejected. Although such a severe condition is uncommon, the potential occurrence of these hazards during earthquakes should be considered during the site selection process. It should be noted that most current seismic design codes are not intended to prevent damage due to collateral seismic hazards. The codes provide minimum required resistance to earthquake ground-shaking without consideration of settlement, slides, subsidence, or faulting in the immediate vicinity of the structure. Following are brief descriptions of these collateral hazards and their potential consequences.

Figure 3-8   Example of surface fault rupture; 1971 San Fernando, California, earthquake (a thrust fault earthquake). (Photo courtesy of the Earthquake Engineering Research Institute.)

Surface Fault Rupture. Surface fault rupture is the abrupt shearing displacement that occurs along a fault that extends to the ground surface when the fault ruptures to cause an earthquake (Figure 3-8). Generally, a fault rupture extends to the ground surface only during earthquakes of magnitude 6 or higher. Surface fault shear displacements typically range from a few inches to a foot or two for a magnitude 6 earthquake, to 10 feet or more for a magnitude 7.5 earthquake. Because fault displacements tend to occur along a relatively narrow area defining the fault zone, large displacements may have catastrophic effects on a structure located directly astride the fault.

Soil Liquefaction. Soil liquefaction is a phenomenon in which a loose granular soil deposit below the ground water table may lose a substantial amount of strength due to earthquake ground shaking. There are many potential adverse consequences of liquefaction, including small building settlements, larger settlements associated with reduction of foundation bearing strength, and large lateral ground displacements that would tend to shear a building apart. An often cited soil liquefaction failure is shown in Figure 3-9.

Soil Differential Compaction. If a site is underlain by loose natural soil deposits, or uncompacted or poorly compacted fill, earthquake ground shaking may cause the soil to be compacted and settle, and differential settlements may occur due to spatial variations in soil properties.

Figure 3-9   Aerial view of leaning apartment houses resulting from soil liquefaction and the behavior of liquefiable soil foundations, Niigata, Japan, earthquake of June 16, 1964. (Photo courtesy of National Oceanic and Atmospheric Administration, National Data Service).

Landsliding. Hillside and sloped sites may be susceptible to seismically-induced landslides. Landslides during earthquakes occur due to horizontal seismic inertia forces induced in the slopes by the ground shaking. Buildings located on slopes, or above or below slopes but close to either the top or the toe of the slope, could be affected by landslides. Landslides having large displacements have devastating effects on a building. An example of a building damaged by a landslide is shown in Figure 3-10.

Inundation. Earthquake-induced flooding at a site can be caused by tsunami (coastal waves caused by some large offshore earthquakes), seiche (waves in bounded bodies of water caused by ground motion), landslides within or entering bodies of water, and the failure of dams. Such hazards are uncommon but need to be considered because of the potentially devastating consequences for sites located in inundated areas. The tilting of a structure caused by tsunami is shown in Figure 3-11.

3.4   OTHER COLLATERAL HAZARDS

In addition to the seismic shaking hazards described in Section 3.2 and the collateral seismic hazards described in Section 3.3, there are other

Figure 3-10   Government Hill School, Anchorage, destroyed by landslide during the magnitude-8.4 Alaska earthquake of 1964. (Photo courtesy of National Oceanic and Atmospheric Administration, National Data Service).

Figure 3-11   Overturned lighthouse at Aonae, Okushiri, from the tsunami following the 1993 Hokkaido-Nansei-Oki earthquake. (Photo courtesy of Yuji Ishiyama, Hokkaido University, Sapporo, Japan)

hazards that are indirectly related to earthquake events. In general, these hazards relate to conditions external to the building site that affect the postearthquake situation, but are outside the control of the building owner and the site selection team. These include: (1) hazards such as vulnerable lifelines (transportation, communication, and utility networks) and hazardous adjacent structures, including buildings close enough to pound against the building that is to be constructed at the site; (2) the storage and distribution of hazardous materials, and (3) postearthquake fires. These other collateral hazards and their potential impacts are described below.

Earthquake-related hazards also include nearby vulnerable lifelines, hazardous adjacent structures, improperly stored hazardous materials, and postearthquake fires.

Vulnerable Lifeline Systems. Earthquake damaged lifeline systems (transportation, communication, and utility networks) may impede the provision of necessary utility functions, or access to the building site in the postearthquake aftermath. Such eventualities are largely outside the control of building designers and managers. The loss of potable water as a result of damage to water storage and distribution systems would make most facilities unusable, as would loss of power due to damage to electric power generation facilities and electric power regional and local distribution lines. Access to certain facilities, such as hospitals, can also be problematic, as for example, in the case of a hospital that is otherwise operable but is inaccessible because of damage to access highways and bridges.

There are numerous examples of transportation lifeline failures during earthquakes and the consequent disruption to facility access. These include freeway bridges damaged during the 1989 Loma Prieta earthquake near San Francisco, and freeway bridges damaged during the 1971 San Fernando and 1994 Northridge earthquakes near Los Angeles. One of the most serious lifeline losses in recent years was the collapse of an upper-deck span on the Oakland-San Francisco Bay bridge during the Loma Prieta earthquake, which resulted in the closure of the bridge for one month for damage repair. This closure impacted the economy on both sides of the San Francisco Bay, because the 250,000 daily users of the bridge had to find alternative routes or postpone the transportation of goods and services, commuting to work, and traveling for other purposes, such as to schools, medical facilities, shopping centers, and other business operations.

Pounding and Hazardous Adjacent Structures. In dense urban settings, there exists the potential for closely spaced buildings to pound against each other (Figure 3-12). Pounding occurs when buildings with differ-

IDENTIFYING AND ASSESSING EARTHQUAKE- RELATED HAZARDS

PIERS CRACKED
FROM POUNDING
AT LEVEL OF
ADJACENT ROOF

Figure 3-12   Photo showing damage caused by the pounding of a 10-story
steel-frame building (with masonry infill walls) against a seven-
story building. Most of the cracking damage to the piers of the
taller building was at the roof line of the shorter building
(Most of the cracking damage to the piers of the taller building
was at the roof line of the shorter building. ATC-20 Training
Slide Set photo)

ent dynamic response characteristics, which are governed by building
stiffness (period of vibration), floor height, and number of stories,
vibrate or sway out of sync when subjected to ground shaking. The
potential for pounding is most acute when the story heights of adjacent
buildings are dissimilar. Pounding has caused severe damage
and even collapse in urban earthquakes, such as during the
magnitude-8.1 earthquake that affected Mexico City in 1985.
Although building codes call attention to this problem,
building designers are often reluctant to provide the necessary space between buildings to eliminate the problem, principally because the required space would reduce available
square footage in the building being developed. Consequently, adequate seismic gaps between buildings are seldom implemented in
densely populated urban areas of seismically hazardous regions of the
United States. In Japan, even with the acute shortage of space in its
largest cities, the problem is taken seriously, with new buildings seldom
built closer than a meter or so from adjacent structures. In suburban or
campus-type site planning in which building sites tend to be much
larger, the problem seldom arises.

Building designers are often reluctant to provide the
necessary space between buildings to eliminate
pounding, principally because the required space would
reduce available square footage in the building being
developed.

Closely spaced buildings in dense urban environments are also subjected to the failure of hazardous adjacent buildings or building components. The problem is most acute if there is an older adjacent building, built to less stringent seismic codes, that is taller than the new building being constructed. Designers should carefully assess neighboring structures and design against possible falling objects from them (e.g., unreinforced parapets, walls, or chimneys). In the 1989 Loma Prieta earthquake, several fatalities were caused when a large portion of an unreinforced masonry building collapsed onto the roof of a lower adjoining building. A typical failure is shown in Figure 3-13.

Storage and Distribution of Hazardous Materials. Hazardous materials, such as stored toxic chemicals in industrial buildings, laboratories, and other facilities, can be extremely dangerous to building occupants and neighboring facilities if released during an earthquake due to the fall and failure of containment vessels. The release of natural or liquefied petroleum gas from earthquake damaged storage or pipeline distribution systems can also be potentially hazardous, not only from the toxic standpoint but because of the potential for postearthquake fires (see below).

Postearthquake Fires. Historically, fires have been one of the most common and damaging hazards associated with earthquakes. Extensive fire damage occurred following the San Francisco earthquake of 1906, and the destruction and life loss in 1923 in Tokyo were largely the result of postearthquake fires (Figure 3-14). More recently, entire neighborhoods were destroyed by fire after the 1995 earthquake in Kobe, Japan.

The potential for postearthquake fires depends on the number and proximity of ignition sources, the availability of fuel, and the fire-fighting capability, which relates to available manpower, available fire-fighting equipment, and available water for fighting fires. Building codes in the United States have extensive provisions ensuring that the materials of construction reduce the fuel content of buildings and that building planning and construction, including the provision of space around buildings for vehicular access and for fire-fighting equipment, provides for the safety of occupants. Modern construction codes have significantly reduced the risk for steel and reinforced concrete buildings, but wood-frame construction, the most dominate type of construction in the United States, remains extremely vulnerable to postearthquake fires because of the flammability of the material. Sources of ignition include overturned gas water heaters and earthquake damaged gas distribution pipelines, which often occurs during moderate and large earthquakes,

Figure 3-13　Photo showing fallen parapets from an earthquake-damaged unreinforced masonry building. (ATC-20 Training Slide Set photo)

together with sparks or fires from damaged electrical distribution lines and other sources. The earthquake aftermath may result in special impediments to fire-fighting that do not exist in the ordinary course of events, including wide-spread and multiple fires, damage to fire stations and equipment, injuries to personnel, impeded access to the building sites, and failures in the water supply system.

Figure 3-14   Photo showing the burning of the Tokyo Police Station following the magnitude-8.3 Tokyo/Kanto earthquake of 1923.

3.5   GUIDANCE FOR ASSESSING EARTHQUAKE-RELATED HAZARDS

The design team, which consists of the architect, structural engineer, geotechnical engineer and possibly others, has the responsibility for advising the owner on important earthquake-related hazards, including those affecting site selection and those that can be reduced in the design and construction process. The owner has final authority on site selection, but likely needs advice on earthquake hazard reduction. All members of the team have roles to play in determining and mitigating earthquake-related hazards for the site. The roles merge and become less rigid depending, for example, on the knowledge, experience, authority, and confidence of the owner and the individuals on the design team.

The assessment of potential earthquake-related hazards should be carried out during the site evaluation process. The evaluation of a site for a new building should consider: (a) zoning restrictions and the local authority's planning restrictions; (b) regional geology and its associated regional seismicity, on a scale that spans, for example, from tens to hundreds of miles, providing information on the regional ground shaking hazard, and locations of historically active faults; (c) site soil conditions, on a scale that spans, for example, from tens to hundreds of yards, providing information for foundation support and the local ground shaking hazard; (d) the earthquake survivability of service utilities, transportation infrastructure, other lifelines, and access for employees; and (e) hazards from outside the property boundary, including unfavorable topography (e.g., potential landslides), the potential for inundation due to tsunami or dam failure, neighboring buildings that are so close that pounding is a potential hazard, adjacent buildings that have potentially hazardous components that may fail and fall on the building, and hazardous building contents. The design team's report to the owner should include the impact of the sum of all these evaluations on the desired performance level attained by the completed building in future earthquakes.

A site evaluation checklist is provided in Figure 3-15. A complete evaluation should address the issues shown in this checklist. This checklist is intended to assist in identifying those issues with which the individuals on the design team should be familiar, and the areas where further consultant help may be necessary.

It is clear from the checklist in Figure 3-15 that a geotechnical engineer, and other specialists, should participate in evaluation of the following seismic-related hazards:

○ ground shaking hazard;

○ seismogeologic hazards that could result, for example, in ground failure beneath the building; and

○ collateral on-site and off-site hazards, such as damage to utilities or transportation infrastructure that results from ground failure and could adversely affect an organization's operations.

Specific guidance on the evaluation of strong ground shaking, the evaluation of collateral seismic hazards, and the evaluation of other collateral hazards follows.

Evaluating the Ground Shaking Hazard

Although ground shaking is the primary hazard affecting building performance at most sites, it is often not explicitly considered in site selection because it does not provide the site with a fatal flaw. That is, the cost to design to a higher, or the maximum, ground shaking level would generally not cause a site to be rejected outright. Nevertheless, if alternative sites are being considered, it is desirable to have a geotechnical engineer evaluate the differences in estimated levels of ground shaking among sites because of the potential influence on project costs. The level of shaking is influenced by the characteristics of the faults in the

SITE EVALUATION CHECKLIST

☐ Is there an active fault on or adjacent to the site?

☐ Will the site geology increase ground shaking?
  ○ Does the site contain unconsolidated or man-made fills?

☐ Is the geology stable?

☐ Is the site susceptible to liquefaction?

☐ Are adjacent up-slope and down-slope soils stable?

☐ Are postearthquake access and egress secure?

☐ Are transportation, communication and utility lifelines unusually vulnerable to disruption and failure?

☐ Are there adjacent land uses that could be hazardous after an earthquake?

☐ Are hazardous materials used or stored in close proximity?

☐ Are building setbacks adequate to prevent pounding from adjacent structures?

☐ Are adjacent structures collapse hazards that might impact your structure?

☐ Is the site subject to inundation from tsunami? seiche? dam failure? flooding?

☐ Are there areas of the site that should be left undeveloped due to:
  ○ Landslide potential?
  ○ Inundation potential?
  ○ High liquefaction potential?
  ○ Expected surface faulting?
  ○ More violent or longer duration ground shaking than the code design values?
  ○ Needed separation from adjacent uses or structures?

☐ Is there adequate space on the site for a safe and "defensible" area of refuge from hazards for building occupants?

☐ Does the site plan increase potential for earthquake-induced landslides by:
  ○ cutting unstable slopes,
  ○ increasing the surface runoff, or
  ○ increasing the soil water content?

Figure 3-15   Site evaluation check list.

site region, the distance of the site from the faults, source-to-site ground motion attenuation characteristics, and site soil conditions. Ground motion attenuation is, in turn, influenced by the source-to-site geology. If alternative sites are located at sufficiently different distances from seismic sources in the same region or are located in different regions, then expected levels of shaking at the sites may be different even if site soil conditions are similar. However, for close sites in the same region, the primary factor causing differences in ground shaking levels is the local soil condition. Resources such as national or state ground shaking maps and site factors, for example in the *NEHRP Recommended Provisions for Seismic Regulations for New Buildings and Other Structures* (BSSC, 2001) and the *International Building Code* (ICC, 2000), enable an experienced geotechnical engineer to make an assessment of differences in expected ground shaking levels among sites.

Ground Shaking Maps and Site Factors

1. The *NEHRP Recommended Provisions for Seismic Regulations for New Buildings and Other Structures* (BSSC, 2001)
2. The *International Building Code* (ICC, 2000)

Evaluating Collateral Seismic Hazards

Guidelines for screening and evaluating potential building sites for collateral seismic hazards (surface fault rupture, soil liquefaction, soil differential compaction, landslide, and inundation) are presented in a number of publications including the FEMA 273 report, NEHRP *Guidelines for the Seismic Rehabilitation of Buildings* (ATC/BSSC, 1997a), the companion FEMA 274 report, *Commentary on the NEHRP Guidelines for the Seismic Rehabilitation of Buildings* (ATC/BSSC, 1997b), the FEMA 356 report, *Prestandard and Commentary for the Seismic Rehabilitation of Buildings* (ASCE, 2000) and the U.S. Army Corps of Engineers publication TI809-04 for the seismic design of buildings (USACE, 1998).

Evaluating Collateral Seismic Hazards

1. The FEMA 273 report, *NEHRP Guidelines for the Seismic Rehabilitation of Buildings* (ATC/BSSC, 1997a),
2. The FEMA 274 report, *Commentary on the NEHRP Guidelines for the Seismic Rehabilitation of Buildings* (ATC/BSSC, 1997b),
3. The FEMA 356 report, *Prestandard and Commentary for the Seismic Rehabilitation of Buildings* (ASCE, 2000)
4. The U.S. Army Corps of Engineers publication TI809-04 for the seismic design of buildings (USACE, 1998).

Surface Fault Rupture. Generally, it is not feasible to design a building to withstand large fault displacements. Sites transected by active faults should generally be avoided unless the probability of faulting during the building life is sufficiently low.

Soil Liquefaction. The assessment of the vulnerability of a site to soil liquefaction must address the hazard severity, the potential effects on the building and utility connections, and any need for design measures to mitigate the hazard. The hazard consequences may range from essentially no adverse effects and no increase in development costs to catastrophic effects that cannot be economically mitigated.

Soil Differential Compaction. In general, seismically-induced soil differential settlements will not be large enough to have a major effect on site development or building design costs. Unusual sites may contain thick layers of uncompacted or poorly compacted fill, where seismic (and static) differential settlements could be large and difficult to predict. Even for such sites, building settlements can be minimized, for example, by using deep pile foundations extending below the fill.

Landsliding. During site selection, the focus should be on identifying unstable or marginally stable hillside slopes that could experience large landslide displacements and require significant cost to mitigate. Slopes having pre-existing active or ancient landslides are especially susceptible to landsliding during future earthquakes.

Tsunami Run-Up

Ziony, J.I., Editor, 1985, *Evaluating Earthquake Hazards in the Los Angeles Region — An Earth-Science Perspective,* U.S. Geological Survey, Professional Paper 1360.

Inundation. During site selection, earthquake-induced flooding should be considered, recognizing that tsunami, seiche, landslides within or entering bodies of water, and the failure of dams are uncommon. Such hazards may exist in coastal areas, near large bodies of water, and in the region downstream from large dams. Guidance on tsunami run-up elevations is available in publications such as the U.S. Geological Survey Professional Paper 1360. Guidance on the potential for landslides within or entering bodies of water should be obtained from a geotechnical engineer. Locations of large dams and the potentially affected downstream area should dam failure occur are typically available from dam regulatory agencies.

Evaluating Other Collateral Hazards

The existence of other collateral hazards that emanate from outside the property boundary, such as neighboring buildings that are so close that pounding is a potential hazard, adjacent buildings that have potentially hazardous components that may fail and fall on the building, and hazardous building contents, should be identified as part of a site selection study if alternative sites are under consideration, and the hazards are evaluated in terms of both the probability of occurrence within a certain period and their consequences. Possible collateral hazards related to a selected site should be identified and procedures for their mitigation should form part of the postearthquake building emergency response plan.

In particular, building owners and the design team responsible for site selection should be familiar with hazardous materials stored and used on the building site, as well as the potential for storage in nearby facilities. Planning for the release of hazardous materials is essential if investigation shows that the building site is vulnerable to such hazards. Similarly, building owners and the design team responsible for site selection should consider the potential for damage caused by adjacent structures, either by pounding or the collapse of nearby hazardous buildings or their components.

The means for reducing lifeline system seismic hazards, which could result in the failure of transportation and utility systems as well as the means for reducing the potential for regional fires following earthquakes, are generally outside the control of the building owner and design team.

3.6 REFERENCES AND FURTHER READING

The following publications are suggested resources for further information.

Guidelines, Pre-Standards, and Codes with Information on Evaluating Ground Shaking and Collateral Hazards

ASCE, 2000, *Prestandard and Commentary for the Seismic Rehabilitation of Buildings*, prepared by the American Society of Civil Engineers; published by the Federal Emergency Management Agency, FEMA 356 Report, Washington, DC.

ATC/BSSC, 1997a, *NEHRP Guidelines for the Seismic Rehabilitation of Buildings*, prepared by the Applied Technology Council (ATC-33 project) for the Building Seismic Safety Council; published by the Federal Emergency Management Agency, FEMA 273 Report, Washington, DC.

ATC/BSSC, 1997b, *Commentary on the NEHRP Guidelines for the Seismic Rehabilitation of Buildings*, prepared by the Applied Technology Council (ATC-33 project) for the Building Seismic Safety Council; published by the Federal Emergency Management Agency, FEMA 274 Report, Washington, DC.

BSSC, 2001, *The 2000 NEHRP Recommended Provisions For New Buildings And Other Structures, Part I, Provisions* and *Part 2, Commentary*, prepared by the Building Seismic Safety Council, published by the Federal Emergency Management Agency, FEMA 368 Report, Washington, DC.

ICC, 2000, *International Building Code*, International Code Council, Falls Church, VA.

USACE, 1998, *Seismic Design of Buildings*, TI-809-04, U.S. Army Corps of Engineers, Washington, DC.

Earthquake Reconnaissance Reports (All Published by the Earthquake Engineering Research Institute: www.eeri.org)

Benuska, L., Ed., 1990, *Loma Prieta Earthquake of October 17, 1989: Reconnaissance Report*, Supplement to *Earthquake Spectra* Volume 6, 450 pp.

Comartin, C.D., Ed., 1995, *The Guam Earthquake of August 8, 1993: Reconnaissance Report*, Supplement to *Earthquake Spectra* Volume 11, 175 pp.

Comartin, C.D., Greene, M., and Tubbesing, S.K., Eds., 1995, *The Hyogo-ken Nanbu Earthquake, January 17,1995: Preliminary Reconnaissance Report*, 116 pp.

Chung, R. Ed., 1995, *Hokkaido-nansei-oki, Japan, Earthquake of July 12, 1993: Reconnaissance Report*, Supplement to *Earthquake Spectra* Volume 11, 166 pp.

Cole, E.E. and Shea, G.H., Eds., 1991, *Costa Rica Earthquake of April 22, 1991: Reconnaissance Report*, Special Supplement B to *Earthquake Spectra* Volume 7, 170 pp.

EERI, 1985, *Impressions of the Guerrero-Michoacan, Mexico, Earthquake of 19 September 1985: A Preliminary Reconnaissance Report*, Published in cooperation with the National Research Council of the National Academy of Sciences, 36 pp.

Hall, J., Ed., 1995, *Northridge Earthquake of January 17, 1994: Reconnaissance Report*, Vol. I, Supplement to *Earthquake Spectra* Volume 11, 523 pp.

Hall, J., Ed., 1996, *Northridge Earthquake of January 17, 1994: Reconnaissance Report*, Vol. II, 280 pp.

Malley, J.O., Shea, G.H., and Gulkan, P. Eds., 1993, *Erzincan, Turkey, Earthquake of March 13, 1992: Reconnaissance Report*, Supplement to *Earthquake Spectra* Volume 9, 210 pp.

Schiff, A.J., Ed., 1991, *Philippines Earthquake of July 16, 1990: Reconnaissance Report*, Special Supplement A to *Earthquake Spectra* Volume 7, 150 pp.

Wyllie, L.A. and Filson, J.R., Eds., 1989, *Armenia Earthquake of December 7, 1988: Reconnaissance Report*, Supplement to *Earthquake Spectra* Volume 5, 175 pp.

PERFORMANCE-BASED ENGINEERING: AN EMERGING CONCEPT IN SEISMIC DESIGN 4

Performance-based seismic design, the focus of this chapter, is a relatively new concept that reflects a natural evolution in engineering design practice. It is based on investigations of building performance in past earthquakes and laboratory research, and is enabled by improvements in analytical tools and computational capabilities. Performance-based seismic design concepts have been made possible by the collective intellect of an interested profession and significant financial resources provided in large part by the federally funded National Earthquake Hazards Reduction Program.

To introduce the subject, we begin with a description of the process by which seismic codes are developed and implemented (Section 4.1), followed by a discussion of the expected performance of new buildings designed in accordance with current seismic codes (Section 4.2). Interestingly enough, as discussed in Section 4.3, currently applied concepts in performance-based seismic design were developed for the rehabilitation of existing buildings, as opposed to the design of new buildings. These concepts, however, apply equally well to new buildings, and model codes for new building seismic design are beginning to adopt and adapt the performance-based concepts created for seismic rehabilitation of existing buildings. As described in Section 4.4, work is also underway to develop next-generation performance-based seismic design guidelines for new and existing buildings.

4.1 SEISMIC DESIGN PROVISIONS IN BUILDING CODES

Building design codes for cities, states, or other jurisdictions throughout the United States are typically based on the adoption and occasional modification of a model building code. Up until the mid-1990s, there were three primary model building code organizations: Building Officials and Code Administrators International, Inc. (BOCA), International Conference of Building Officials (ICBO), and Southern Building Code Congress International, Inc. (SBCCI). In 1994, these three organizations united to found the International Code Council (ICC), a non-profit organization dedicated to developing a single set of comprehensive and coordinated national model construction codes. The first code published by ICC was the *2000 International Building Code* (IBC; ICC, 2000).

Building code adoption is a complicated process, especially in regions with significant exposure to natural hazards such as earthquake, wind, or flood. In some earthquake-prone regions of the United States, the seismic design provisions outlined in the 2000 IBC have not been adopted. Instead, the provisions of the *Uniform Building Code* (UBC), the model building code published by IBCO from 1949 through 1997 (ICBO, 1997), are still used. The seismic provisions in the UBC are based primarily on the provisions contained in the Structural Engineers Association of California (SEAOC) *Recommended Lateral Force Requirements and Commentary*, known as the *Blue Book* and published from 1959 through 1999 (SEAOC, 1999). In addition, the 1997 UBC relies on the provisions contained in the 1994 edition of the *NEHRP Recommended Provisions for Seismic Regulations for New Buildings* (BSSC, 1995), while the 2000 IBC relies on the more recent 1997 edition of the *NEHRP Recommended Provisions for Seismic Regulations for New Buildings and Other Structures* (BSSC, 1998).

The *NEHRP Provisions* have been published regularly since the National Earthquake Hazards Reduction Program (NEHRP) was created in 1978 as a response to Congress passing P.L. 95-124, the Earthquake Hazards Reduction Act of 1977. The Federal Emergency Management Agency (FEMA) was mandated to implement P.L. 95-124 and NEHRP, and the Building Seismic Safety Council (BSSC) was formed to provide a broad consensus mechanism for regularly updating the *NEHRP Recommended Provisions for Seismic Regulations for New Buildings* (hereinafter referred to as the *NEHRP Recommended Provisions*), first published in 1978 by the Applied Technology Council as *Tentative Provisions for the Development of Seismic Regulations for Buildings*, (ATC-03 Report; ATC, 1978). The most recent version of the *NEHRP Recommended Provisions* is the 2000 edition (BSSC, 2001) and a 2003 edition is currently in development, as discussed later in this chapter.

The remainder of this chapter explores seismic design issues related to current building codes, specifically the intent of current codes with respect to the performance of structural and nonstructural building systems. The current codes include the 2000 IBC and the 1997 UBC (and the *NEHRP Recommended Provisions* and SEAOC *Blue Book* on which they rely), as these are most commonly used, although it should be noted that a few jurisdictions have adopted the recently published National Fire Protection Agency (NFPA) 5000 *Building Construction and Safety Code* (NFPA, 2003) in conjunction with the American Society of Civil Engineers (ASCE, 2002) ASCE 7-02 publication, *Minimum Design Loads for Buildings and Other Structures*, for earthquake loading requirements.

Performance-based engineering, an emerging design tool for managing seismic risk, and the impact that the emergence of performance-based design strategies will have on future buildings and their seismic performance, discussed later in this chapter.

4.2 EXPECTED PERFORMANCE WHEN DESIGNING TO CURRENT CODES

The basic intent of current seismic design provisions is best summarized by the SEAOC *Recommended Lateral Force Requirements and Commentary* (SEAOC, 1999), which states:

> These Requirements provide minimum standards for use in building design regulation to maintain public safety in the extreme ground shaking likely to occur during an earthquake. These Requirements are primarily intended to safeguard against major failures and loss of life, not to limit damage, maintain functions, or provide for easy repair.

In other words, current seismic design codes are essentially aimed at the preservation of life and safety for the benefit of the community. The recommended provisions express expectations and provide no guarantees; they assume that there may be damage to a building as a result of an earthquake. For example, the SEAOC *Recommended Lateral Force Requirements and Commentary* includes a general set of performance statements to qualify the nature of expected damage, as follows:

> Structures designed in accordance with these recommendations should, in general, be able to:
>
> ○ Resist a minor level of earthquake ground motion without damage
>
> ○ Resist a moderate level of earthquake ground motion without structural damage, but possibly experience some nonstructural damage.
>
> ○ Resist a major level of earthquake ground motion having an intensity equal to the strongest either experienced or forecast for the building site without collapse, but possibly with some structural as well as nonstructural damage.

> It is expected that structural damage, even in a major design level earthquake, will be limited to a repairable level for most structures that meet these Requirements. In some instances, damage may not be economical to repair. The level of damage depends upon a

DESIGN CONSIDERATION

Current seismic design codes are essentially aimed at the preservation of life and safety for the benefit of the community.

number of factors, including the intensity and duration of ground shaking, structure configuration, type of lateral force resisting system, materials used in the construction, and construction workmanship.

Codes do not provide the designer with the difference in performance between different structural systems.

Designers use codes as a resource, as they provide minimum acceptable consensus standards. Codes provide no guidance on the selection of materials and systems, rather only criteria for their use once selected. Codes also do not provide the designer with the difference in performance between systems; for example, the difference between the stiffness of shear walls and frames and the importance of this characteristic for the overall seismic performance of the building. Lastly, codes do not discuss that the use of some structural systems will result in more nonstructural damage than others, even though the structural systems perform equally well in resisting the earthquake forces. The following sub-sections describe the expected performance of structural and nonstructural components, respectively.

Expected Performance of Structural Components

Current seismic design provisions for non-essential facilities are intended to provide resistance to collapse in a major earthquake (typically the design ground motion). Resistance to collapse means that the structure may have lost a substantial amount of its original lateral stiffness and strength, but the gravity-load-bearing elements still function and provide some margin of safety against collapse.

As mentioned earlier, current seismic design provisions for non-essential facilities are intended to provide life safety, i.e., no damage in a minor earthquake, limited structural damage in a moderate earthquake, and resistance to collapse in a major earthquake (typically the design ground motion). Resistance to collapse means that the structure may have lost a substantial amount of its original lateral stiffness and strength, but the gravity-load-bearing elements still function and provide some margin of safety against collapse. The structure may have permanent lateral offset and some elements of the seismic-force resisting system may exhibit substantial cracking, spalling, yielding, buckling, and localized failure. Following a major earthquake, the structure is not safe for continued occupancy until repairs are done. Shaking associated with strong aftershocks could threaten the stability of the structure. Repair to a structure in this state is expected to be feasible, however it may not be economically attractive to do so. Section 4.3 includes further discussion of the seismic behavior of specific structural systems in the context of describing performance-based design objectives.

Figure 4-1   Photo of lights set into a fixed ceiling system that shook loose during an earthquake and are hanging from their conduits. (ATC-20 Training Slide Set photo)

Expected Performance of Nonstructural Components

While current seismic design provisions provide minimum structural performance standards in terms of resistance to collapse, they typically do not address performance of nonstructural components, such as room partitions, filing cabinets and book cases, hung lighting and ceilings, entryway canopies, and stairwells; nor do they address performance of mechanical, electrical, or plumbing systems including fire sprinklers, heating and air conditioning equipment or ductwork, and electrical panels or transformers. The vast majority of damage and resulting loss of building functionality during recent damaging earthquakes has been the result of damage to nonstructural components and systems (Figure 4-1). Many building owners have been surprised when a building withstands the effects of a moderate earthquake from a structural perspective, but is still rendered inoperable from a nonstructural standpoint.

While current seismic design provisions provide minimum structural performance standards in terms of resistance to collapse, they typically do not address performance of nonstructural components.

Current seismic design provisions typically require that nonstructural components be secured so as to not present a falling hazard; however, these components can still be severely damaged such that they can not function. Loss of electric power, breaks in water supply and sewer out-

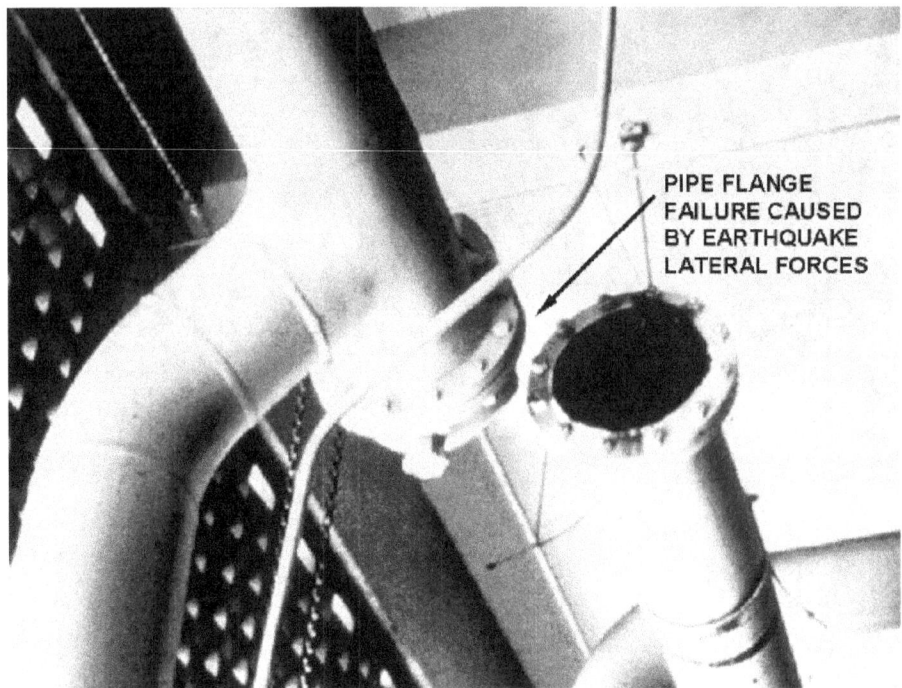

PIPE FLANGE
FAILURE CAUSED
BY EARTHQUAKE
LATERAL FORCES

Figure 4-2   Photo of pipe flange failure caused by earthquake lateral forces. (ATC-20 Training Slide Set photo)

flow lines (see Figure 4-2), or non-functioning heating or air conditioning will render a building unusable by tenants. Breaks in fire sprinklers will cause flooding within all or part of a building, soaked carpets and walls, inundated files and records, and electrical shorts or failures in electrical equipment and computers. Other examples of nonstructural damage that can be expected in a code-compliant building subjected to strong ground shaking include extensive cracking in cladding, glazing, partitions, and chimneys; broken light fixtures; racked doors; and dropped ceiling tiles. Section 4.3 includes further discussion of the seismic behavior of specific nonstructural systems in the context of describing performance-based design objectives.

4.3 CURRENT SPECIFICATIONS FOR PERFORMANCE-BASED SEISMIC DESIGN

As described earlier, an important yet emerging concept in the successful implementation of seismic risk management strategies is the application of performance-based seismic design approaches. The primary function of performance-based seismic design is the ability to achieve, through analytical means, a building design that will reliably perform in a prescribed manner under one or more seismic hazard condi-

DESIGN CONSIDERATION

The primary function of performance-based seismic design is the ability to achieve, through analytical means, a building design that will reliably perform in a prescribed manner under one or more seismic hazard conditions.

tions. The fact that alternative levels of building performance are being defined and can be chosen as performance objectives is a relatively new development in seismic design. Some of its origins lie in studies of building performance during recent earthquakes, in which owners of buildings that suffered hundreds of thousands of dollars in damage were surprised to learn that the buildings met the intent of the life-safety provisions of the seismic code under which they were designed, since no one was killed or seriously injured. Out of these experiences came the realization that design professionals need to be more explicit about what "design to code" represents and what seismic design in general can and can not accomplish. At the same time, studies of damaged buildings, together with laboratory research and computer analyses, have led to a much more sophisticated understanding of building response under the range of earthquake ground motion that can be expected to occur.

This section describes some of the key concepts of performance-based seismic design. These concepts have emerged from a series of studies, funded by FEMA, that focused on the development of performance-based seismic design guidelines for existing buildings. The first study, published in the FEMA 237 report, *Seismic Rehabilitation of Buildings – Phase I: Issues Identification and Resolution* (ATC, 1992), identified and resolved a wide variety of scope, format, socio-economic, and detailed technical issues that needed to be considered during the subsequent development of practical guidelines for the seismic rehabilitation of buildings. During that initial study, the concept of performance goals was introduced, effectively commencing the move toward performance-based seismic design. The follow-on study, an $8-million FEMA-funded effort carried out jointly by the Applied Technology Council, the Building Seismic Safety Council, and the American Society of Civil Engineers, resulted in the formulation of detailed guidelines, written for practicing structural engineers and building officials, that specify the means to use performance-based design concepts to rehabilitate existing buildings to improve their seismic resistance. The final set of products of that effort consists of three documents: FEMA 273, NEHRP *Guidelines for the Seismic Rehabilitation of Buildings* (ATC/BSSC, 1997a); FEMA 274, *NEHRP Commentary for the Guidelines for Seismic Rehabilitation of Buildings* (ATC/ BSSC, 1997b); and FEMA 276, *Example Applications of the NEHRP Guidelines for the Seismic Rehabilitation of Buildings* (ATC, 1997). In order to speed the implementation of the FEMA 273 *Guidelines* in structural engineering practice, the *Guidelines* were converted, with funding from FEMA, to a *Prestandard and Commentary for the Seismic Rehabilitation of Buildings* (FEMA 356) by the American Society of Civil Engineers

(ASCE, 2000). The conversion process maintained the performance levels and performance descriptions, as described below, and other concepts developed for the FEMA 273 *Guidelines*.

Midway through the long-term FEMA effort to develop the FEMA 273 *Guidelines* and FEMA 356 *Prestandard*, the Structural Engineers Association of California (SEAOC) developed *Vision 2000, Performance Based Seismic Engineering of Buildings*, which describes a framework for performance based seismic design of new buildings. At about the same time, the Applied Technology Council developed the ATC-40 report, *Seismic Evaluation and Rehabilitation of Concrete Buildings* (ATC, 1996b), a detailed procedures manual for the seismic evaluation and rehabilitation of concrete buildings using performance-based seismic design concepts. All of the documents described above are important, state-of-the-art resources for the structural engineering community.

The application of performance-based seismic design can be highly technical, and requires that the design engineer have a good understanding of seismic hazards, and the dynamic and inelastic behavior of buildings and materials. Unlike the application of building codes, performance-based seismic design is not typically prescriptive in nature, and often requires significantly more detailed building analysis than might otherwise be required. However, as discussed earlier, the advantages of performance-based seismic design in the development of an overall risk management plan is usually worth the extra effort spent by the design team. It is the challenge of the design team to convey this level of importance to the owner and the owner's representatives.

Building Performance Objectives

A fundamental concept behind the implementation of performance-based seismic design is the development of a consensus set of performance objectives. The performance objectives describe the intended performance of the building (e.g., in terms of life safety, levels of acceptable damage, and post-earthquake functionality) when subjected to an earthquake hazard of a defined intensity (e.g., a maximum credible event or an event with a certain return period). As earthquake intensity increases, building performance generally decreases. The goal of specifying a performance objective is to achieve a reliable estimate of

Engineering Applications for Performance-Based Seismic Design

○ ATC-40, *Seismic Evaluation and Retrofit of Concrete Buildings*, Applied Technology Council, 1996b.

○ FEMA-273, *NEHRP Guidelines for the Seismic Rehabilitation of Buildings*, (ATC/BSSC, 1997a).

○ FEMA-274, *Commentary on NEHRP Guidelines for the Seismic Rehabilitation of Buildings*, (ATC/BSSC, 1997b).

○ FEMA-356, *Prestandard and Commentary for the Seismic Rehabilitation of Buildings* (ASCE, 2000).

○ SEAOC, *Vision 2000: Performance Based Seismic Engineering of Buildings*, Structural Engineers Association of California, 1995.

Building Performance Objective
Intended performance level in combination with a specified seismic shaking level.

Table 4-1   Performance Objectives (Adapted from FEMA 356 (ASCE, 2000))

		Target Building Performance Levels*			
		Operational Performance Level (1-A)	Immediate Occupancy Performance Level (1-B)	Life Safety Performance Level (3-C)	Collapse Prevention Performance Level (5-E)
Earthquake Hazard Level (ground motions having a specified probability of being exceeded in a 50-year period)	50%/50 year	a	b	c	d
	20%/50 year	e	f	g	h
	BSE-1 (10%/50 year)	i	j	k	l
	BSE-2 (2%/50 year)	m	n	o	p

*Alpha-numeric identifiers in parentheses defined in Table 4-2

Notes:
1. Each cell in the above matrix represents a discrete Rehabilitation Objective
2. Three specific Rehabilitation Objectives are defined in FEMA 356:

Basic Safety Objective	= cells k + p
Enhanced Objectives	= cells k + p + any of a, e, i, b, f, j, or n
Limited Objectives	= cell k alone, or cell p alone
Limited Objectives	= cells c, g, d, h, l

performance under one or more earthquake hazard scenarios. A representation of different performance objectives is shown in Table 4-1, which is taken from FEMA 356 *Prestandard and Commentary for Seismic Rehabilitation of Buildings* ((ASCE, 2000), a prestandard for performance-based seismic rehabilitation of existing buildings. Although FEMA 356 pertains to seismic rehabilitation, the same concepts apply to new design.

As shown in Table 4-1, FEMA 356 defines two basic earthquake hazard levels – Basic Safety Earthquake 1 (BSE-1, corresponding to 475-year return period event, or ground motions having a 10% probability of being exceeded in 50 years) and Basic Safety Earthquake 2 (BSE-2 corresponding to 2475-year return period event, or ground motions having a 2% probability of being exceeded in 50 years). The Basic Safety Objective (BSO) is then defined as meeting the target building performance level of Life Safety for BSE-1, and the target building performance level of Collapse Prevention for BSE-2 (cells k and p in Table 4-1). FEMA 356 states,

> The BSO is intended to approximate the earthquake risk to life safety traditionally considered acceptable in the United States.

Buildings meeting the BSO are expected to experience little damage from relatively frequent, moderate earthquakes, but significantly more damage and potential economic loss from the most severe and infrequent earthquakes that could affect them.

Performance objectives higher than the BSO are defined as Enhanced Objectives, which are achieved by designing for target building performance levels greater than those of the BSO at either the BSE-1 or BSE-2 hazard levels or by designing for the target building performance levels of the BSO using an earthquake hazard level that exceeds either the BSE-1 or BSE-2 (see Table 4-1 notes). The possible combinations of target building performance and earthquake hazard level corresponding to design for an enhanced performance objective are limitless – the goal is simply to provide building performance better than that intended by the BSO and mandated in most current design codes.

Note also that certain cells of the matrix (Table 4-1) are referred to as Limited Objectives. While this lower performance objective may be applicable to certain partial or reduced seismic rehabilitation designs, it does not apply to new design as it falls below current code standards.

Building Performance Levels

Building performance can be described qualitatively in terms of the:

- safety afforded building occupants, during and after an earthquake.
- cost and feasibility of restoring the building to pre-earthquake conditions.
- length of time the building is removed from service to conduct repairs.
- economic, architectural, or historic impacts on the community at large.

These performance characteristics will be directly related to the extent of damage sustained by the building during a damaging earthquake. As shown in Table 4-1, FEMA 356 defines four basic Target Building Performance Levels, which differ only slightly in terminology from the four levels described in FEMA 369, *The 2000 NEHRP Recommended Provisions for New Buildings and Other Structures, Part 2: Commentary* (BSSC, 2001). These performance levels, illustrated graphically in Figure 4-3, are:

- **Operational Level:** The lowest level of overall damage to the building. The structure will retain nearly all of its pre-earthquake strength and stiffness. Expected damage includes minor cracking

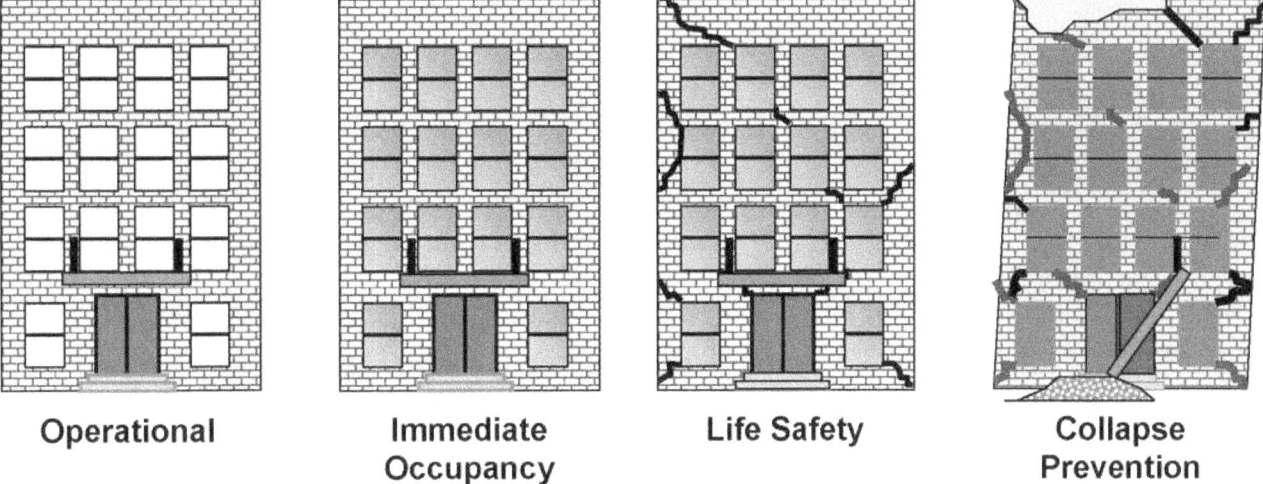

| Operational | Immediate Occupancy | Life Safety | Collapse Prevention |

Figure 4-3   Graphic illustration of Operational, Immediate Occupancy, Life-Safety, and Collapse Prevention Performance Levels. (Courtesy of R. Hamburger)

of facades, partitions, and ceilings, as well as structural elements. All mechanical, electrical, plumbing, and other systems necessary for normal operation of the buildings are expected to be functional, possibly from standby sources. Negligible damage to nonstructural components is expected. Under very low levels of earthquake ground motion, most buildings should be able to meet or exceed this performance level. Typically, however, it will not be economically practical to design for this level of performance under severe levels of ground shaking, except for buildings that house essential services.

○ **Immediate Occupancy Level:** Overall damage to the building is light. Damage to the structural systems is similar to the Operational Performance Level; however, somewhat more damage to nonstructural systems is expected. Nonstructural components such as cladding and ceilings, and mechanical and electrical components remain secured; however, repair and cleanup may be needed. It is expected that utilities necessary for normal function of all systems will not be available, although those necessary for life safety systems would be provided. Many building owners may wish to achieve this level of performance when the building is subjected to moderate levels of earthquake ground motion. In addition, some owners may desire such performance for very important buildings, under severe levels of earthquake ground shaking. This level provides most of the protection obtained under the Operational Building Performance Level, without the associated cost of providing standby utilities and performing rigorous seismic qualification to validate equipment performance.

○ **Life Safety Level:** Structural and nonstructural damage is significant. The building may lose a substantial amount of its pre-earthquake lateral strength and stiffness, but the gravity-load-bearing elements function. Out-of-plane wall failures and tipping of parapets are not expected, but there will be some permanent drift and select elements of the lateral-force resisting system may have substantial cracking, spalling, yielding, and buckling. Nonstructural components are secured and not presenting a falling hazard, but many architectural, mechanical, and electrical systems are damaged. The building may not be safe for continued occupancy until repairs are done. Repair of the structure is feasible, but it may not be economically attractive to do so. This performance level is generally the basis for the intent of code compliance.

○ **Collapse Prevention Level** or **Near Collapse Level:** The structure sustains severe damage. The lateral-force resisting system loses most of its pre-earthquake strength and stiffness. Load-bearing columns and walls function, but the building is near collapse. Substantial degradation of structural elements occurs, including extensive cracking and spalling of masonry and concrete elements, and buckling and fracture of steel elements. Infills and unbraced parapets may fail and exits may be blocked. The building has large permanent drifts. Nonstructural components experience substantial damage and may be falling hazards. The building is unsafe for occupancy. Repair and restoration is probably not practically achievable. This building performance level has been selected as the basis for mandatory seismic rehabilitation ordinances enacted by some municipalities, as it results in mitigation of the most severe life-safety hazards at relatively low cost.

Building performance levels typically comprise a structural performance level that describes the limiting damage state of the structural systems, plus a nonstructural performance level that describes the limiting damage state of the nonstructural systems and components. Table 4-2, from FEMA 356, illustrates this concept. A Target Building Performance Level is designated by the number corresponding to the Structural Performance Level (identified as S-1 through S-6) and the letter corresponding to the Nonstructural Performance Level (identified as N-A through N-E). Note that in Tables 4-1 and 4-2, the four Target Building Performance Levels discussed above are each designated as follows.

Table 4-2   Target Building Performance Levels and Ranges (ASCE, 2000)

Nonstructural Performance Levels	Structural Performance Levels and Ranges					
	S-1 Immediate Occupancy	S-2 Damage Control Range	S-3 Life Safety	S-4 Limited Safety Range	S-5 Collapse Prevention	S-6 Not Considered
N-A Operational	Operational (1-A)	2-A	NR[1]	NR[1]	NR[1]	NR[1]
N-B Immediate Occupancy	Immediate Occupancy (1-B)	2-B	3-B	NR[1]	NR[1]	NR[1]
N-C Life Safety	1-C	2-C	Life Safety (3-C)	4-C	5-C	6-C
N-D Hazards Reduced	NR[1]	2-D	3-D	4-D	5-D	6-D
N-E Not Considered	NR[1]	NR[1]	NR[1]	4-E	Collapse Prevention (5-E)	NR[1]

Notes:
1. NR = Not Recommended

○ **Operational Level** (1-A): Immediate Occupancy Structural Performance Level (S-1) plus Operational Nonstructural Performance Level (N-A).

○ **Immediate Occupancy Level** (1-B): Immediate Occupancy Structural Performance Level (S-1) plus Immediate Occupancy Nonstructural Performance Level (N-B).

○ **Life Safety Level** (3-C): Life Safety Structural Performance Level (S-3) plus Life Safety Nonstructural Performance Level (N-C).

○ **Collapse Prevention Level** (5-E): Collapse Prevention Structural Performance Level (S-5) plus Not Considered Nonstructural Performance Level (N-E).

Note also that in Table 4-2, there are several combinations of structural and nonstructural performance levels that are not recommended for rehabilitation; the same lack of recommendation applies to new design. The six structural performance levels and five nonstructural performance levels are described in the following subsections.

Structural Performance Levels

For the rehabilitation of existing buildings, the Structural Performance Levels most commonly used are the Immediate Occupancy Level, the

Life Safety Level, and the Collapse Prevention Level (S-1, S-3, and S-5, respectively, in Table 4-2). These levels are discrete points on a continuous scale describing the building's expected performance or, alternatively, how much damage, economic loss, and disruption may occur following the design earthquake. Intermediate levels are often used to assist in quantifying the continuous scale. For example, Table 4-2 lists Structural Performance Levels of Damage Control Range (S-2), Limited Safety Range (S-4), and Not Considered (S-6).

Structural Performance Levels relate to the limiting damage states for common elements of the building's lateral force resisting systems. Table 4-3 and Table 4-4, taken from FEMA 356, provide descriptions of the damage associated with the three Structural Performance Levels of Collapse Prevention, Life Safety, and Immediate Occupancy for specific types of horizontal (Table 4-3) and vertical (Table 4-4) structural elements and systems.

Nonstructural Performance Levels

The four Nonstructural Performance Levels most commonly used are the Operational Level, the Immediate Occupancy Level, the Life Safety Level, and the Hazards Reduced Level (N-A, N-B, N-C, and N-D, respectively, in Table 4-2). Table 4-2 also includes the additional performance level of Not Considered (N-E). Nonstructural components addressed

Table 4-3   Structural Performance Levels and Damage — Horizontal Elements (From FEMA 356)

Element	Performance Levels		
	Collapse Prevention	Life Safety	Immediate Occupancy
Metal Deck Diaphragms	Large distortion with buckling of some units and tearing of many welds and seam attachments.	Some localized failure of welded connections of deck to framing and between panels. Minor local buckling of deck.	Connections between deck units and framing intact. Minor distortions.
Wood Diaphragms	Large permanent distortion with partial withdrawal of nails and extensive splitting of elements.	Some splitting at connections. Loosening of sheathing. Observable withdrawal of fasteners. Splitting of framing and sheathing.	No observable loosening or withdrawal of fasteners. No splitting of sheathing or framing.
Concrete Diaphragms	Extensive crushing and observable offset across many cracks.	Extensive cracking (< 1/4" width). Local crushing and spalling.	Distributed hairline cracking. Some minor cracks of larger size (< 1/8" width).
Precast Diaphragms	Connections between units fail. Units shift relative to each other. Crushing and spalling at joints.	Extensive cracking (< 1/4" width). Local crushing and spalling.	Some minor cracking along joints.

Table 4-4 Structural Performance Levels and Damage[1] — Vertical Elements (from FEMA 356)

Elements	Type	Structural Performance Levels		
		Collapse Prevention	Life Safety	Immediate Occupancy
Concrete Frames	Primary	Extensive cracking and hinge formation in ductile elements. Limited cracking and/or splice failure in some nonductile columns. Severe damage in short columns.	Extensive damage to beams. Spalling of cover and shear cracking (< 1/8" width) for ductile columns. Minor spalling in nonductile columns. Joint cracks < 1/8" wide.	Minor hairline cracking. Limited yielding possible at a few locations. No crushing (strains below 0.003).
	Secondary	Extensive spalling in columns (limited shortening) and beams. Severe joint damage. Some reinforcing buckled.	Extensive cracking and hinge formation in ductile elements. Limited cracking and/or splice failure in some nonductile columns. Severe damage in short columns.	Minor spalling in a few places in ductile columns and beams. Flexural cracking in beams and columns. Shear cracking in joints < 1/16" width.
	Drift[2]	4% transient or permanent	2% transient; 1% permanent	1% transient; negligible permanent
Steel Moment Frames	Primary	Extensive distortion of beams and column panels. Many fractures at moment connections, but shear connections remain intact.	Hinges form. Local buckling of some beam elements. Severe joint distortion; isolated moment connection fractures, but shear connections remain intact. A few elements may experience partial fracture.	Minor local yielding at a few places. No fractures. Minor buckling or observable permanent distortion of members.
	Secondary	Same as primary.	Extensive distortion of beams and column panels. Many fractures at moment connections, but shear connections remain intact.	Same as primary.
	Drift[2]	5% transient or permanent	2.5% transient; 1% permanent	0.7% transient; negligible permanent
Braced Steel Frames	Primary	Extensive yielding and buckling of braces. Many braces and their connections may fail.	Many braces yield or buckle but do not totally fail. Many connections may fail.	Minor yielding or buckling of braces.
	Secondary	Same as primary.	Same as primary.	Same as primary.
	Drift[2]	2% transient or permanent	1.5% transient; 0.5% permanent	0.5% transient; negligible permanent

Concrete Walls	Primary	Major flexural and shear cracks and voids. Sliding at joints. Extensive crushing and buckling of reinforcement. Failure around openings. Severe boundary element damage. Coupling beams shattered and virtually disintegrated.	Some boundary element distress, including limited buckling of reinforcement. Some sliding at joints. Damage around openings. Some crushing and flexural cracking. Coupling beams: extensive shear and flexural cracks; some crushing, but concrete generally remains in place.	Minor hairline cracking of walls, < 1/16" wide. Coupling beams experience cracking < 1/8" width.
	Secondary	Panels shattered and virtually disintegrated.	Major flexural and shear cracks. Sliding at joints. Extensive crushing. Failure around openings. Severe boundary element damage. Coupling beams shattered and virtually disintegrated.	Minor hairline cracking of walls. Some evidence of sliding at construction joints. Coupling beams experience cracks < 1/8" width. Minor spalling.
	Drift[2]	2% transient or permanent	1% transient; 0.5% permanent	0.5% transient; negligible permanent
Unreinforced Masonry Infill Walls[3]	Primary	Extensive cracking and crushing; portions of face course shed.	Extensive cracking and some crushing but wall remains in place. No falling units. Extensive crushing and spalling of veneers at corners of openings.	Minor (<1/8" width) cracking of masonry infills and veneers. Minor spalling in veneers at a few corner openings.
	Secondary	Extensive crushing and shattering; some walls dislodge.	Same as primary.	Same as primary.
	Drift[2]	0.6% transient or permanent	0.5% transient; 0.3% permanent	0.1% transient; negligible permanent
Unreinforced Masonry (Noninfill) Walls	Primary	Extensive cracking; face course and veneer may peel off. Noticeable in- plane and out-of-plane offsets.	Extensive cracking. Noticeable in-plane offsets of masonry and minor out-of-plane offsets.	Minor (< 1/8" width) cracking of veneers. Minor spalling in veneers at a few corner openings. No observable out-of- plane offsets.
	Secondary	Nonbearing panels dislodge.	Same as primary.	Same as primary.
	Drift[2]	1% transient or permanent	0.6% transient; 0.6% permanent	0.3% transient; 0.3% permanent

Reinforced Masonry Walls	Primary	Crushing; extensive cracking. Damage around openings and at corners. Some fallen units.	Extensive cracking (< 1/4") distributed throughout wall. Some isolated crushing.	Minor (< 1/8" width) cracking. No out-of-plane offsets.
	Secondary	Panels shattered and virtually disintegrated.	Crushing; extensive cracking; damage around openings and at corners; some fallen units.	Same as primary.
	Drift[2]	1.5% transient or permanent	0.6% transient; 0.6% permanent	0.2% transient; 0.2% permanent
Wood Stud Walls	Primary	Connections loose. Nails partially withdrawn. Some splitting of members and panels. Veneers dislodged.	Moderate loosening of connections and minor splitting of members.	Distributed minor hairline cracking of gypsum and plaster veneers.
	Secondary	Sheathing sheared off. Let-in braces fractured and buckled. Framing split and fractured.	Connections loose. Nails partially withdrawn. Some splitting of members and panels.	Same as primary.
	Drift[2]	3% transient or permanent	2% transient; 1% permanent	1% transient; 0.25% permanent
Precast Concrete Connections	Primary	Some connection failures but no elements dislodged.	Local crushing and spalling at connections, but no gross failure of connections.	Minor working at connections; cracks < 1/16" width at connections.
	Secondary	Same as primary.	Some connection failures but no elements dislodged.	Minor crushing and spalling at connections.
Foundations	General	Major settlement and tilting.	Total settlements < 6" and differential settlements < 1/2" in 30 ft.	Minor settlement and negligible tilting.

Notes:

1. The damage states indicated in this table are provided to allow an understanding of the severity of damage that may be sustained by various structural elements when present in structures meeting the definitions of the Structural Performance Levels. These damage states are not intended for use in post- earthquake evaluation of damage nor for judging the safety of, or required level of repair to, a structure following an earthquake.

2. The drift values, differential settlements, and similar quantities indicated in these tables are not intended to be used as acceptance criteria for evaluating the acceptability of a rehabilitation design in accordance with the analysis procedures provided in these Guidelines; rather, they are indicative of the range of drift that typical structures containing the indicated structural elements may undergo when responding within the various performance levels. Drift control of a rehabilitated structure may often be governed by the requirements to protect nonstructural components. Acceptable levels of foundation settlement or movement are highly dependent on the construction of the superstructure. The values indicated are intended to be qualitative descriptions of the approximate behavior of structures meeting the indicated levels.

3. For limiting damage to frame elements of infilled frames, refer to the rows for concrete or steel frames.

by these performance levels include architectural components (e.g., partitions, exterior cladding, and ceilings) and mechanical and electrical components (e.g., HVAC systems, plumbing, fire suppression systems, and lighting). Occupant contents and furnishings (such as inventory and computers) are often included as well.

Nonstructural Performance Levels relate to the limiting damage states for common elements of the building's architectural features, utility systems, and contents and other equipment. Tables 4-5, 4-6, and 4-7 taken from FEMA 356, provide descriptions of the damage associated with the four Nonstructural Performance Levels of Hazards Reduced, Life Safety, Immediate Occupancy, and Operational for specific types of architectural components (Table 4-5); mechanical, electrical, and plumbing system components (Table 4-6); and contents (Table 4-7).

4.4 IMPACT OF PERFORMANCE-BASED STRATEGIES ON FUTURE DESIGN CODES

While current design codes explicitly require life safety design for only a single level of ground motion, it is expected that future design codes will provide engineers with the necessary guidelines to design and construct buildings that meet a number of performance criteria when subjected to earthquake ground motion of differing severity. The current (2000) version of the *NEHRP Recommended Provisions for Seismic Regulations for New Buildings and Other Structures* (BSSC, 2001) has initiated a move towards incorporating performance-based strategies through the use of three "Seismic Use Groups." These groups are categorized based on the occupancy of the structures and the relative consequences of earthquake-induced damage to the structures as follows:

○ Group III structures are essential facilities required for postearthquake recovery, and those structures that contain significant amounts of hazardous materials. An example is a medical facility with emergency treatment facilities.

○ Group II structures are those having a large number of occupants and those where the occupants ability to exit is restrained. An example is an elementary school.

○ Group I structures are all other structures, basically those with a lesser life hazard only insofar as there is expected to be fewer occupants in the structures and the structures are lower and/or smaller. An example is a low-rise commercial office building.

The *2000 NEHRP Recommended Provisions for Seismic Regulations of New Buildings and Other Structures* specify progressively more conservative

Table 4-5 Nonstructural Performance Levels and Damage — Architectural Components (from FEMA 356)

Component	Nonstructural Performance Levels			
	Hazards Reduced Level	Life Safety	Immediate Occupancy	Operational
Cladding	Severe damage to connections and cladding. Many panels loosened.	Severe distortion in connections. Distributed cracking, bending, crushing, and spalling of cladding elements. Some fracturing of cladding, but panels do not fall.	Connections yield; minor cracks (< 1/16" width) or bending in cladding.	Connections yield; minor cracks (< 1/16" width) or bending in cladding.
Glazing	General shattered glass and distorted frames. Widespread falling hazards.	Extensive cracked glass; little broken glass.	Some cracked panes; none broken.	Some cracked panes; none broken
Partitions	Severe racking and damage in many cases.	Distributed damage; some severe cracking, crushing, and racking in some areas.	Cracking to about 1/16" width at openings. Minor crushing and cracking at corners.	Cracking to about 1/16" width at openings. Minor crushing and cracking at corners.
Ceilings	Most ceilings damaged. Light suspended ceilings dropped. Severe cracking in hard ceilings.	Extensive damage. Dropped suspended ceiling tiles. Moderate cracking in hard ceilings.	Minor damage. Some suspended ceiling tiles disrupted. A few panels dropped. Minor cracking in hard ceilings.	Generally negligible damage. Isolated suspended panel dislocations, or cracks in hard ceilings.
Parapets and Ornamentation	Extensive damage; some fall in nonoccupied areas.	Extensive damage; some falling in nonoccupied areas.	Minor damage.	Minor damage.
Canopies & Marquees	Extensive distortion.	Moderate distortion.	Minor damage.	Minor damage.
Chimneys & Stacks	Extensive damage. No collapse.	Extensive damage. No collapse.	Minor cracking.	Negligible damage.
Stairs & Fire Escapes	Extensive racking. Loss of use.	Some racking and cracking of slabs, usable.	Minor damage.	Negligible damage.
Light Fixtures	Extensive damage. Falling hazards occur.	Many broken light fixtures. Falling hazards generally avoided in heavier fixtures (> 20 pounds).	Minor damage. Some pendant lights broken.	Negligible damage.
Doors	Distributed damage. Many racked and jammed doors.	Distributed damage. Some racked and jammed doors.	Minor damage. Doors operable.	Minor damage. Doors operable.

System/ Component	Nonstructural Performance Levels			
	Hazards Reduced	Life Safety	Immediate Occupancy	Operational
Elevators	Elevators out of service; counterweights off rails.	Elevators out of service; counterweights do not dislodge.	Elevators operable; can be started when power available.	Elevators operate.
HVAC Equipment	Most units do not operate; many slide or overturn; some suspended units fall.	Units shift on supports, rupturing attached ducting, piping, and conduit, but do not fall.	Units are secure and most operate if power and other required utilities are available.	Units are secure and operate; emergency power and other utilities provided, if required.
Ducts	Ducts break loose of equipment and louvers; some supports fail; some ducts fall.	Minor damage at joints of sections and attachment to equipment; some supports damaged, but ducts do not fall.	Minor damage at joints, but ducts remain serviceable.	Negligible damage.
Piping	Some lines rupture. Some supports fail. Some piping falls.	Minor damage at joints, with some leakage. Some supports damaged, but systems remain suspended.	Minor leaks develop at a few joints.	Negligible damage.
Fire Sprinkler Systems	Many sprinkler heads damaged by collapsing ceilings. Leaks develop at couplings. Some branch lines fail.	Some sprinkler heads damaged by swaying ceilings. Leaks develop at some couplings.	Minor leakage at a few heads or pipe joints. System remains operable.	Negligible damage.
Fire Alarm Systems	Ceiling mounted sensors damaged. System nonfunctional.	May not function.	System is functional.	System is functional.
Emergency Lighting	Some lights fall. Power may not be available.	System is functional.	System is functional.	System is functional.
Electrical Distribution Equipment	Units slide and/or overturn, rupturing attached conduit. Uninterruptable Power Source systems fail. Diesel generators do not start.	Units shift on supports and may not operate. Generators provided for emergency power start; utility service lost.	Units are secure and generally operable. Emergency generators start, but may not be adequate to service all power requirements.	Units are functional. Emergency power is provided, as needed.
Plumbing	Some fixtures broken; lines broken; mains disrupted at source.	Some fixtures broken, lines broken; mains disrupted at source.	Fixtures and lines serviceable; however, utility service may not be available.	System is functional. On-site water supply provided, if required.

Table 4-7   Nonstructural Performance Levels and Damage—Contents (from FEMA 356)

| Contents Type | Nonstructural Performance Levels | | | |
	Hazards Reduced	Life Safety	Immediate Occupancy	Operational
Computer Systems	Units roll and overturn, disconnect cables. Raised access floors collapse.	Units shift and may disconnect cables, but do not overturn. Power not available.	Units secure and remain connected. Power may not be available to operate, and minor internal damage may occur.	Units undamaged and operable; power available.
Manufacturing Equipment	Units slide and overturn; utilities disconnected. Heavy units require reconnection and realignment. Sensitive equipment may not be functional.	Units slide, but do not overturn; utilities not available; some realignment required to operate.	Units secure, and most operable if power and utilities available.	Units secure and operable; power and utilities available.
Desktop Equipment	Units slide off desks.	Some equipment slides off desks.	Some equipment slides off desks.	Equipment secured to desks and operable.
File Cabinets	Cabinets overturn and spill contents.	Drawers slide open; cabinets tip.	Drawers slide open, but cabinets do not tip.	Drawers slide open, but cabinets do not tip.
Book Shelves	Shelves overturn and spill contents.	Books slide off shelves.	Books slide on shelves.	Books remain on shelves.
Hazardous Materials	Severe damage; no large quantity of material released.	Minor damage; occasional materials spilled; gaseous materials contained.	Negligible damage; materials contained.	Negligible damage; materials contained.
Art Objects	Objects damaged by falling, water, dust.	Objects damaged by falling, water, dust.	Some objects may be damaged by falling.	Objects undamaged.

strength, drift control, system selection, and detailing requirements for structures contained in the three groups, in order to attain minimum levels of earthquake performance suitable to the individual occupancies. The design criteria for each group are intended to produce specific types of performance in design earthquake events, based on the importance of reducing structural damage and improving life safety. Figure 4-4, taken from the *Commentary to the 2000 NEHRP Provisions*, illustrates this concept.

Next-Generation Performance-Based Seismic Design Guidelines

Incorporation of performance-based engineering concepts in future design codes will also be aided by a major effort recently initiated by the Applied Technology Council (ATC) with funding from the Federal Emergency Management Agency (FEMA). The project, known as

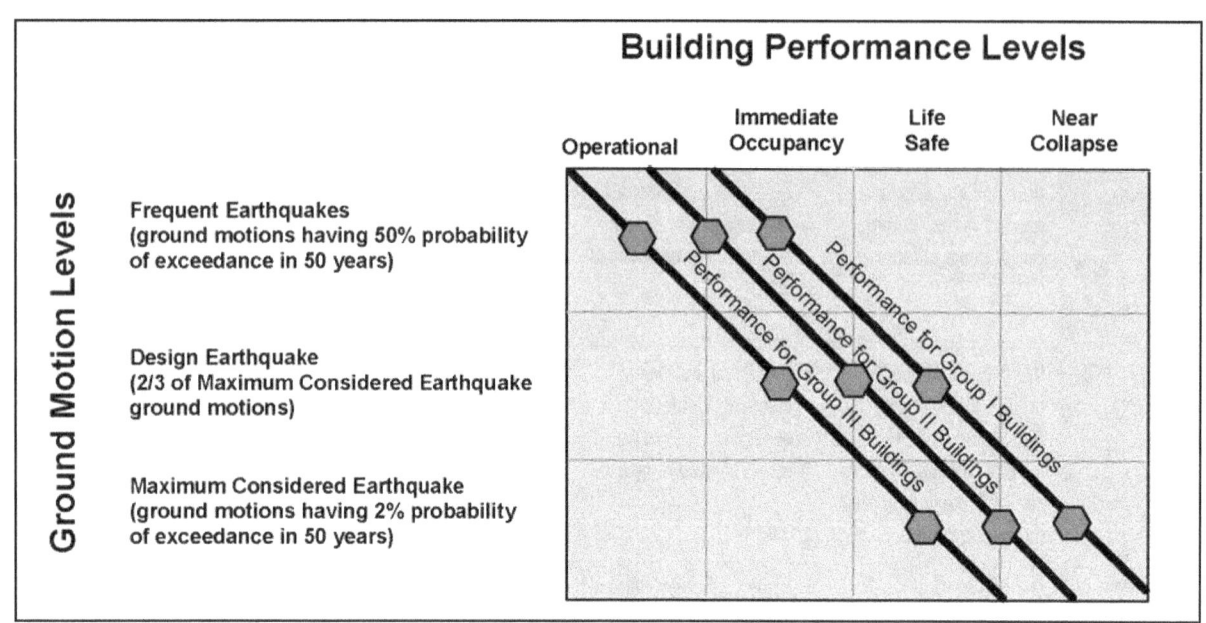

Building Performance Levels

Ground Motion Levels

Frequent Earthquakes (ground motions having 50% probability of exceedance in 50 years)

Design Earthquake (2/3 of Maximum Considered Earthquake ground motions)

Maximum Considered Earthquake (ground motions having 2% probability of exceedance in 50 years)

Operational | Immediate Occupancy | Life Safe | Near Collapse

Performance for Group I Buildings

Performance for Group II Buildings

Performance for Group III Buildings

Figure 4-4   Expected building performance. (from BSSC, 2001)

ATC-58, Development of Performance-Based Seismic Design Guidelines, is currently in Phase I, Project Initiation and Performance Characterization. FEMA has provided funding for Phase I of a planned multi-year program to develop the guidelines, following the general approach outlined in FEMA 349, *Action Plan for Performance Based Seismic Design* (EERI, 2000). It is expected that the successful development of the guidelines will require a multi-year effort entailing financial and technical participation from the four NEHRP agencies as well as private industry.

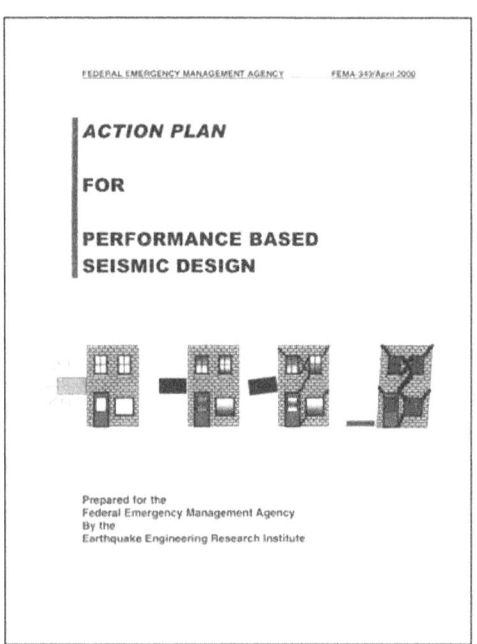

FEMA 349 identifies six products essential to the creation and implementation of comprehensive, acceptable Performance-Based Seismic Design Guidelines:

1. A *Program Management Plan* that incorporates a broadly based oversight group to shepherd and promote the development of the Guidelines (over an extended period of time, say up to 10 years), and an education and implementation strategy to facilitate the use of the Guidelines.

2. *Structural Performance Products* that characterize building performance, specify how to evaluate a building's performance capability for a specified level of seismic hazard and with a defined reliability or level of confidence, and provide guidance on how to design a structure to provide desired performance (with defined reliability).

3. *Nonstructural Performance Products* that provide engineers with the capability to evaluate and design nonstructural components, such as partitions, piping, HVAC (heating, ventilation, and air conditioning) equipment, with the goal of ensuring that such components will provide desired performance (with defined reliability).

4. *Risk Management Products* that provide methodologies for calculating the benefits of designing to various performance objectives and to make rational economic choices about the levels of performance desired, the levels of confidence desired, and the comparative costs to reach those levels.

5. *Performance-Based Seismic Design Guidelines* that provide methodology and criteria for design professionals, material suppliers, and equipment manufacturers to implement performance-based design procedures.

6. A *Stakeholders Guide* that explains performance-based seismic design to nontechnical audiences, including building owners, managers, and lending institutions.

4.5 GUIDANCE FOR DESIGN PROFESSIONALS

It is clear that performance-based strategies will be included in future seismic design codes. Regardless of when this actually occurs, design professionals can utilize the information in this document, as well as those referenced below, to provide owners and managers with a much clearer picture of what may be expected in terms of damage, downtime, and occupant safety for a given building under various intensities of ground motion.

When communicating with building owner representatives during the development of seismic performance criteria for a new building, it would be useful to:

1. Explain the concepts of performance-based seismic design using the concepts and materials provided in this document and the references cited.

2. Help the owner to determine if a performance level higher than life safety is needed for the design earthquake; if so, use these materials and the references cited to assist the owner in developing a design that would be accepted by the governing regulatory agency (e.g., local building department).

4.6 REFERENCES AND FURTHER READING

ASCE, 2000, *Prestandard and Commentary for the Seismic Rehabilitation of Buildings,* prepared by the American Society of Civil Engineers, published by the Federal Emergency Management Agency, *FEMA 356 Report,* Washington, DC.

ATC, 1996b, *Seismic Evaluation and Retrofit of Concrete Buildings,* Volumes 1 and 2, Applied Technology Council, *ATC-40 Report,* Redwood City, California.

ATC, 2004a (in preparation), *Improvement of Inelastic Seismic Analysis Procedures,* prepared by the Applied Technology Council for the Federal Emergency Management Agency, FEMA 440, Washington, D.C.

ATC, 2004b (in preparation), *Program Plan for Development of Performance-Based Seismic Design Guidelines,* prepared by the Applied Technology Council for the Federal Emergency Management Agency, FEMA 445, Washington, D.C.

ATC, 2004c (in preparation), *Characterization of Seismic Performance for Buildings,* prepared by the Applied Technology Council for the Federal Emergency Management Agency, FEMA 446, Washington, D.C.

ATC/BSSC, 1997a, *NEHRP Guidelines for the Seismic Rehabilitation of Buildings,* prepared by the Applied Technology Council (ATC-33 project) for the Building Seismic Safety Council, published by the Federal Emergency Management Agency, *FEMA 273 Report,* Washington, DC.

ATC/BSSC, 1997b, *NEHRP Commentary on the Guidelines for the Seismic Rehabilitation of Buildings,* prepared by the Applied Technology Council (ATC-33 project) for the Building Seismic Safety Council, published by the Federal Emergency Management Agency, *FEMA 274 Report,* Washington, DC.

BSSC, 2001, *NEHRP Recommended Provisions for Seismic Regulations for New Buildings and Other Structures, Part I: Provisions,* and *Part II, Commentary,* 2000 Edition, prepared by the Building Seismic Safety Council, published by the Federal Emergency Management Agency, Publications, *FEMA 368 Report* and *FEMA 369 Report,* Washington, DC.

SEAOC, 1995, *Vision 2000: Performance-Based Seismic Engineering of Buildings,* Structural Engineers Association of California, Sacramento, California.

SEAOC, 1999, *Recommended Lateral Force Requirements and Commentary*, prepared by the Structural Engineers Association of California, published by the International Conference of Building Officials, Whittier, California.

PERFORMANCE-BASED ENGINEERING: AN EMERGING CONCEPT IN SEISMIC DESIGN

IMPROVING PERFORMANCE TO REDUCE SEISMIC RISK 5

5.1 INTRODUCTION

Improving performance to reduce seismic risk is a multi-faceted issue that requires consideration of a broad range of factors. Previous chapters in this document have introduced and described the overarching concept of seismic risk management (Chapter 2) and two of the fundamental factors affecting improved seismic performance: consideration of the seismic hazards affecting the site (Chapter 3); and consideration of the desired seismic performance of structural and nonstructural components for the range of earthquakes of concern (Chapter 4).

This chapter identifies and addresses related seismic design issues that are fundamentally important to improved seismic performance, regardless of the occupancy type:

- selection of the structural materials and systems (Section 5.2);

- selection of the architectural/structural configuration (Section 5.3);

- consideration of the expected performance of nonstructural components, including ceilings, partitions, heating, ventilation, and air condition equipment (HVAC), piping and other utility systems, and cladding (Section 5.4);

- cost analysis, including consideration of both the benefits and costs of improved seismic performance (Sections 5.5 through 5.7);

- and quality control during the construction process (Section 5.8).

Considerable attention is given to the quantification of benefits and costs of improved seismic performance, given the underlying importance of cost considerations. Benefits include reduced direct capital losses and reduced indirect losses, which are related to the time that a given building is operationally out of service. Cost issues are demonstrated through several means, including the use of (1) graphics showing the relationship between the cost of various options for improving seismic performance versus the resulting benefits; and (2) case studies demonstrating best practices in earthquake engineering.

The Chapter concludes with a set of general recommendations for improving seismic performance during the seismic design and construction process, regardless of occupancy type. The subsequent six chapters focus on seismic design and performance issues related to specific occupancy types: commercial office buildings (Chapter 6): retail commercial facilities (Chapter 7); light manufacturing facilities (Chapter 8); healthcare facilities (Chapter 9); local schools, kindergarten through grade 12 (Chapter 10); and higher education (university) facilities (Chapter 11).

5.2 SELECTION OF STRUCTURAL MATERIALS AND SYSTEMS

An earthquake has no knowledge of building function, but uncovers weaknesses in the building that are the result of errors or deficiencies in its design and construction. However, variations in design and construction will affect its response, perhaps significantly, and to the extent that these variations are determined by the occupancy, then each building type tends to have some unique seismic design determinants. A building that uses a moment–frame structure will have a different ground motion response than a building that uses shear walls; the frame structure is more flexible, so it will experience lower earthquake forces, but it will deflect more than the shear wall structure, and this increased motion may cause more damage to nonstructural components such as partitions and ceilings. The shear wall building will be much stiffer but this will attract more force: the building will deflect less but will experience higher accelerations and this will affect acceleration-sensitive components such as air conditioning equipment and heavy tanks.

These structural and nonstructural system characteristics can be deduced from the information in the seismic code, but the code is not a design guide and gives no direct guidance on the different performance characteristics of available systems or how to select an appropriate structural system for a specific site or building type.

Table 5-1 illustrates the seismic performance of common structural systems, both old and new, and gives some guidance as to the applicability of systems and critical design characteristics for good performance. The different structural performance characteristics mean that their selection must be matched to the specific building type and its architecture. Table 5-1 summarizes a great deal of information and is intended only to illustrate the point that structural systems vary in their performance. The table is not intended as the definitive tool for system selection; this requires extensive knowledge, experience and analysis.

Table 5-2 shows structural system selections that are appropriate for different site conditions, for different occupancies and various building functions. For example, an important aspect of the building site is that

Table 5-1   Seismic Performance of Structural Systems (adapted from Elsesser, 1992)

SUMMARY OF SEISMIC PERFORMANCE OF STRUCTURAL SYSTEMS			
Structural System	**Earthquake Performance**	**Specific Building Performance and Energy Absorption**	**General Comments**
Wood Frame	San Francisco, 1906 Alaska 1964 Other Earthquakes Variable to *Good*	○ San Francisco Buildings performed reasonably well even though not detailed. ○ Energy Absorption is excellent	○ Connection details are critical. ○ Configuration is significant
Unreinforced Masonry Wall	San Francisco, 1906 Santa Barbara, 1925 Long Beach, 1933 Los Angeles, 1994 Variable to *Poor*	○ Unreinforced masonry has performed poorly when *not* tied together. ○ Energy absorption is good if system integrity is maintained.	○ Continuity and ties between walls and diaphragm is essential.
Steel Frame with Masonry Infill	San Francisco, 1906 Variable to *Good*	○ San Francisco buildings performed very well. ○ Energy absorption is excellent.	○ Building form must be uniform, relatively small bay *sizes.*
Reinforced Concrete Wall	San Francisco, 1957 Alaska, 1964 Japan 1966 Los Angeles, 1994 Variable to *Poor*	○ Buildings in Alaska, San Francisco and Japan performed poorly with spandrel and pier failure ○ Brittle system	○ Proportion of spandrel and piers is critical, detail for ductility and shear.
Steel Brace	San Francisco, 1906 Taft, 1952 Los Angeles, 1994 Variable	○ Major braced systems performed well. ○ Minor bracing and tension braces performed poorly.	○ Details and proportions are critical.
Steel Moment Frame	Los Angeles, 1971 Japan, 1978 Los Angeles, 1994 ? *Good*	○ Los Angeles and Japanese buildings 1971/78 performed well. ○ Energy absorption is excellent. ○ Los Angeles 1994, mixed performance.	○ Both conventional and ductile frame have performed well if designed for drift.
Concrete Shear Wall	Caracas, 1965 Alaska, 1964 Los Angeles, 1971 Algeria, 1980 *Variable*	○ Poor performance with discontinuous walls. ○ Uneven energy absorption.	○ *Configuration* is *critical,* soft story or L-shape with torsion have produced failures.
Precast Concrete	Alaska, 1964 Bulgaria, 1978 San Francisco, 1980 Los Angeles, 1994 Variable to *Poor*	○ Poor performance in 1964, 1978, 1980, 1994	○ Details for continuity are critical ○ *Ductility* must be achieved
Reinforced Concrete Ductile Moment Frame	Los Angeles, 1971 ? *Good*	○ Good performance in 1971, Los Angeles ○ System will crack ○ Energy absorption is good. ○ Mixed performance in 1994 Los Angeles	○ Details *critical.*

Table 5-2   Structural Systems for Site Conditions and Occupancy Types (from Elsesser, 1992)

STRUCTURAL SYSTEMS FOR SITE CONDITIONS AND OCCUPANCY TYPES			
Site Conditions	"Soft" Site (Long Period)	Use rigid building with short period	Shear Wall · Steel Brace · Eccentric Braced Frame
	Distant Site (short period)	Use rigid building with short period	
	"Hard" Site (Short Period)	Use flexible building with long period	Ductile Moment Frame · Base Isolation
	Poor Soils (Pile Supported)	Use lightweight rigid building	Steel Braced Frame · Steel Tube Frame
Occupancy	High-Tech (labs, computers, hospitals)	Use ductile rigid systems for damage control	Eccentric Braced Frame · Dual Wall / Ductile Moment Frame · Eccentric Braced Frame
	Office Buildings	Open Plan	Steel Ductile Moment Frame · Steel Braced Frame · Eccentric Braced Frame
	Residential	Cellular Spaces	Concrete Shear Wall · Steel Braced Frame

a major structure must be "de-tuned," that is, designed such that its fundamental period differs sufficiently from that of the ground so that dangerous resonance and force amplification are not induced. Thus, for a soft, long-period site; it is appropriate to use a rigid short period structural system; this need in turn must be related to other requirements of occupancy and function.

Table 5-2 also illustrates that structures must be matched to the building's use. For example, a concrete shear wall structure is appropriate for an apartment house because the strong cross walls are an economical way to provide the necessary seismic resistance and, at the same time, provide good acoustics between the apartments. While the purpose of Table 5-2 is to illustrate the way in which structural systems may be matched to the site condition and building design and use, the table is not intended as the definitive tool for system selection; this also requires extensive knowledge, experience, and analysis.

5.3 SELECTION OF THE ARCHITECTURAL CONFIGURATION

The architectural configuration—the building's size, proportions and three-dimensional form—plays a large role in determining seismic performance. This is because the configuration largely determines the distribution of earthquake forces, that is, the relative size and nature of the forces as they work their way through the building. A good configuration will provide for a balanced force distribution, both in plan and section, so that the earthquake forces are carried directly and easily back to the foundations. A poor configuration results in stress concentrations and torsion, which at their worst are dangerous.

Configuration problems have long been identified, primarily as the result of extensive observation of building performance in earthquakes. However, many of the problem configurations arise because they are useful and efficient in supporting the functional needs of the building or accommodating site constraints. The design task is to create configuration alternatives that satisfy both the architectural needs and provide for structural safety and economy. This requires that the architect and engineer must cooperate from the outset of the design process: first to arrive at an appropriate structural system to satisfy building needs, and then to negotiate detailed design alternatives that avoid, or reduce, the impact of potential problem configurations.

Seismic codes now have provisions intended to deal with configuration problems. However, the code approach is to accept the problems and attempt to solve them either by increasing design forces, or requiring a more sophisticated analysis. Neither of these approaches is satisfactory, for they do not remove the problem. In addition, many of the code provisions apply only to buildings that are five stories or over 65 feet in height, which leaves a large number of buildings unregulated by the code. The problem can only be solved by design and not by a prescriptive code.

Design solutions for a soft first story condition that the architect and engineer might explore together include (see Figure 5-1):

○ The architectural implications of eliminating it (which solves the structural problem);

○ Alternative framing designs, such as increasing the number of columns or increasing the system stiffness by changing the design, to alleviate the stiffness discrepancy between the first and adjacent floors; and

○ Adding bracing at the end of line of columns (if the site constraints permit this).

A more general problem is the increasing unpredictability of building response as the architectural/structural configuration increasingly deviates from an ideal symmetrical form. This has serious implications for Performance Based Design, which depends for its effectiveness on the ability of the engineer to predict structural performance.

Tables 5-3 and 5-4 illustrate the above points by identifying the common configuration problems- termed "irregularities" that are dealt with in the seismic code. These are classified as vertical or plan irregularities. The tables show a diagram of each condition, illustrates the failure pattern and describes its effects. The designations and numbers of the conditions are identical to the code: the diagrams are not contained in the code but are interpretations of the descriptions of each condition that the code defines.

5.4 CONSIDERATION OF NONSTRUCTURAL COMPONENT PERFORMANCE

As discussed in Section 4.2, the majority of the damage that has resulted in building closure following recent U.S. earthquakes has been the result of damage to nonstructural components and systems. A building designed to current seismic regulations may perform well structurally in a moderate earthquake, but be rendered nonfunctional due to nonstructural damage.

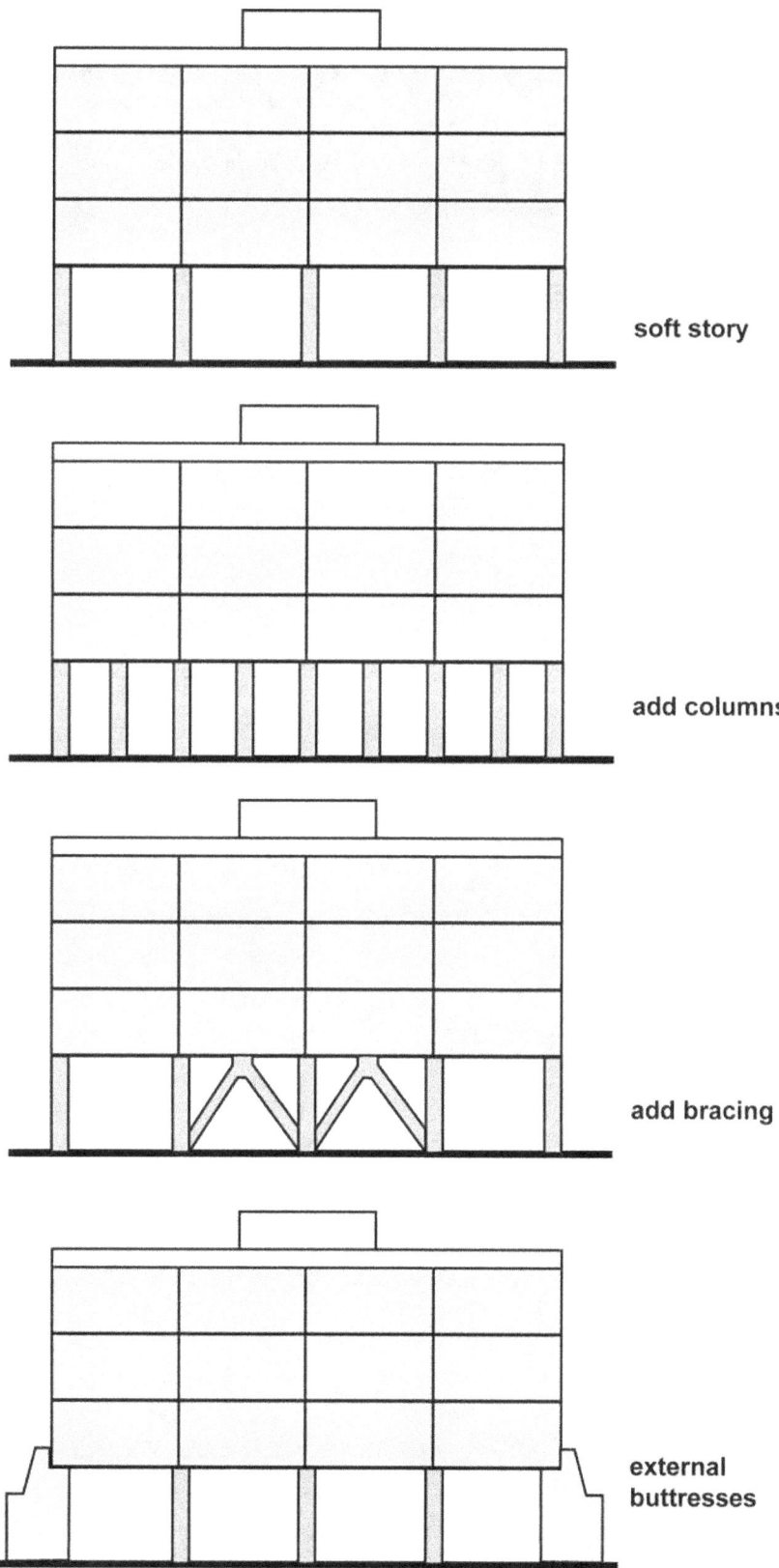

soft story

add columns

add bracing

external
buttresses

Figure 5-1   Example design solutions for addressing soft story condition.

Table 5-3   Vertical Irregularities, Resulting Failure Patterns, and Performance Implications

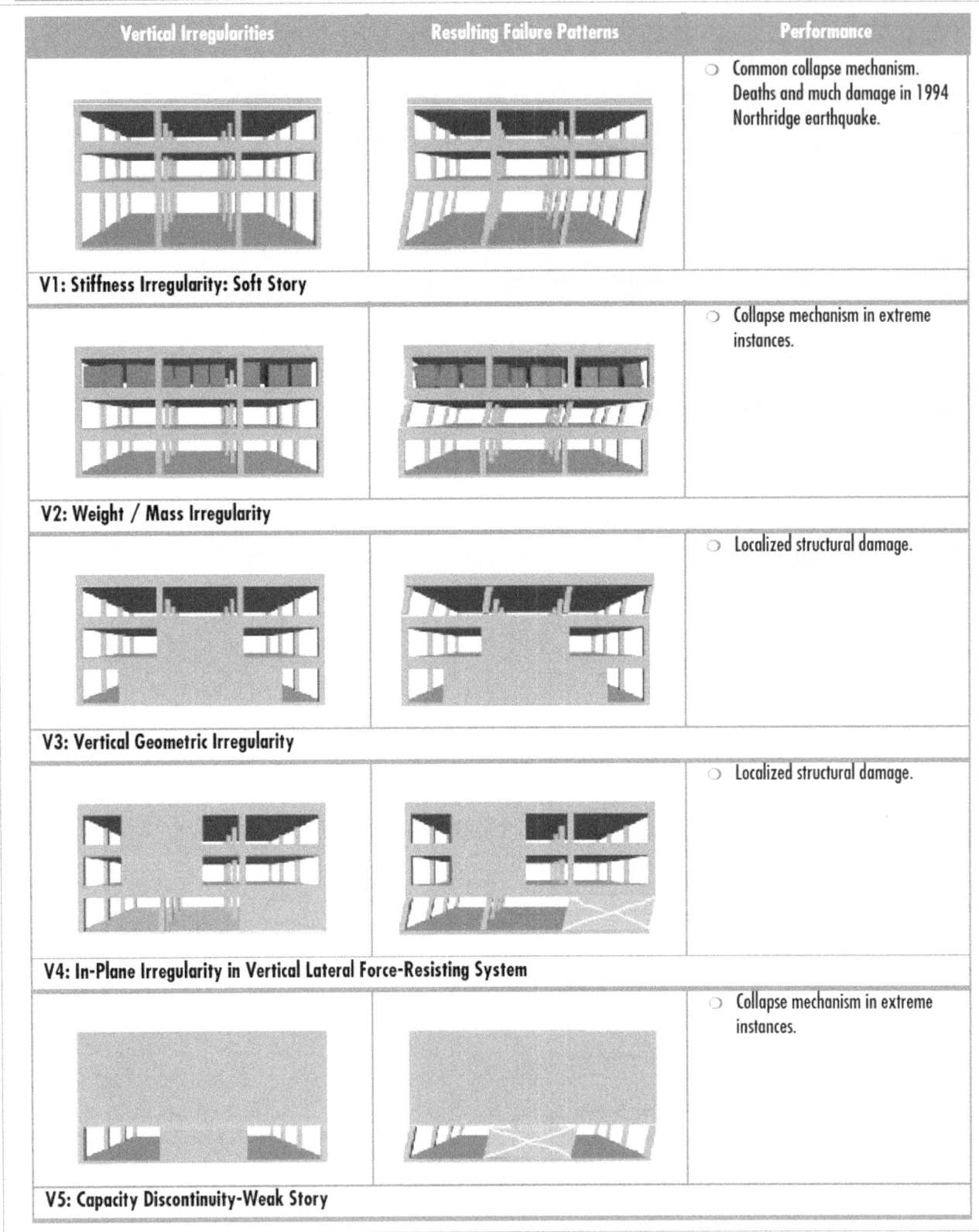

Vertical Irregularities	Resulting Failure Patterns	Performance
V1: Stiffness Irregularity: Soft Story		○ Common collapse mechanism. Deaths and much damage in 1994 Northridge earthquake.
V2: Weight / Mass Irregularity		○ Collapse mechanism in extreme instances.
V3: Vertical Geometric Irregularity		○ Localized structural damage.
V4: In-Plane Irregularity in Vertical Lateral Force-Resisting System		○ Localized structural damage.
V5: Capacity Discontinuity-Weak Story		○ Collapse mechanism in extreme instances.

Table 5-4   Plan Irregularities, Resulting Failure Patterns, and Performance Implications

Plan Irregularities	Resulting Failure Patterns	Performance
		○ Localized damage. ○ Collapse mechanism in extreme instances.
P1: Torsional Irregularity: Unbalanced Resistance		
		○ Localized damage to diaphragms and attached elements. ○ Collapse mechanism in extreme instances in large buildings.
P2: Reentrant Corners		
		○ Localized damage to diaphragms and attached elements.
P3: Diaphragm Eccentricity and Cut-outs		
		○ Collapse mechanism in extreme instances.
P4: Out-of-Plane Offsets: Discontinuous Shear Walls		
		○ Leads to torsion and instability, localized damage.
P5: Nonparallel Lateral Force-Resisting Systems		

Nonstructural components may also, however, influence structural performance in response to ground shaking. Structural analysis to meet code requirements assumes a bare structure. Nonstructural components that are attached to the structure, and heavy contents, depending on their location, may introduce torsional forces. Characteristic examples of structural/nonstructural interaction are as follows:

○ Heavy masonry partitions that are rigidly attached to columns and under floor slabs, can, if asymmetrically located, introduce localized stiffness and create stress concentrations and torsional forces. A particular form of this condition, that has caused significant structural damage, is when short column conditions are created by the insertion of partial masonry walls between columns. The addition of such partial walls after the building completion is often treated as a minor remodel that is not seen to require engineering analysis. The result is that the shortened columns have high relative stiffness, attract a large percentage of the earthquake forces, and fail (Figure 5-2).

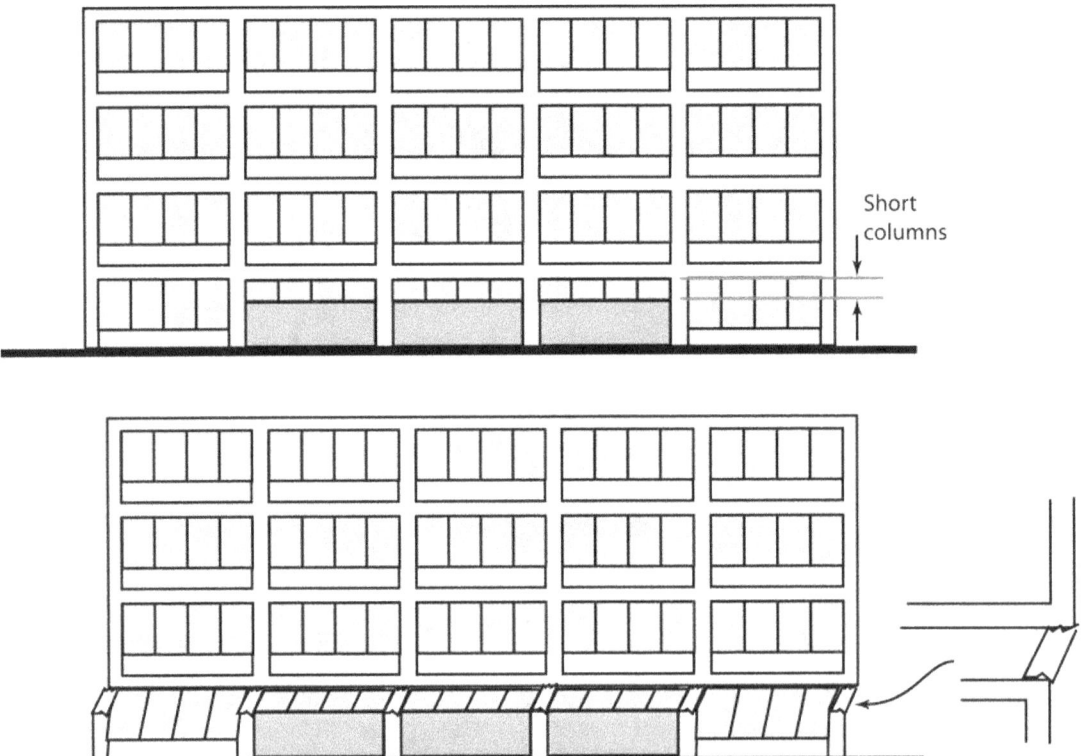

Figure 5-2   Elevation views of building with short columns between first and second floors. Upper sketch show the building in an unshaken state; lower sketch shows damage mechanism under earthquake lateral loading.

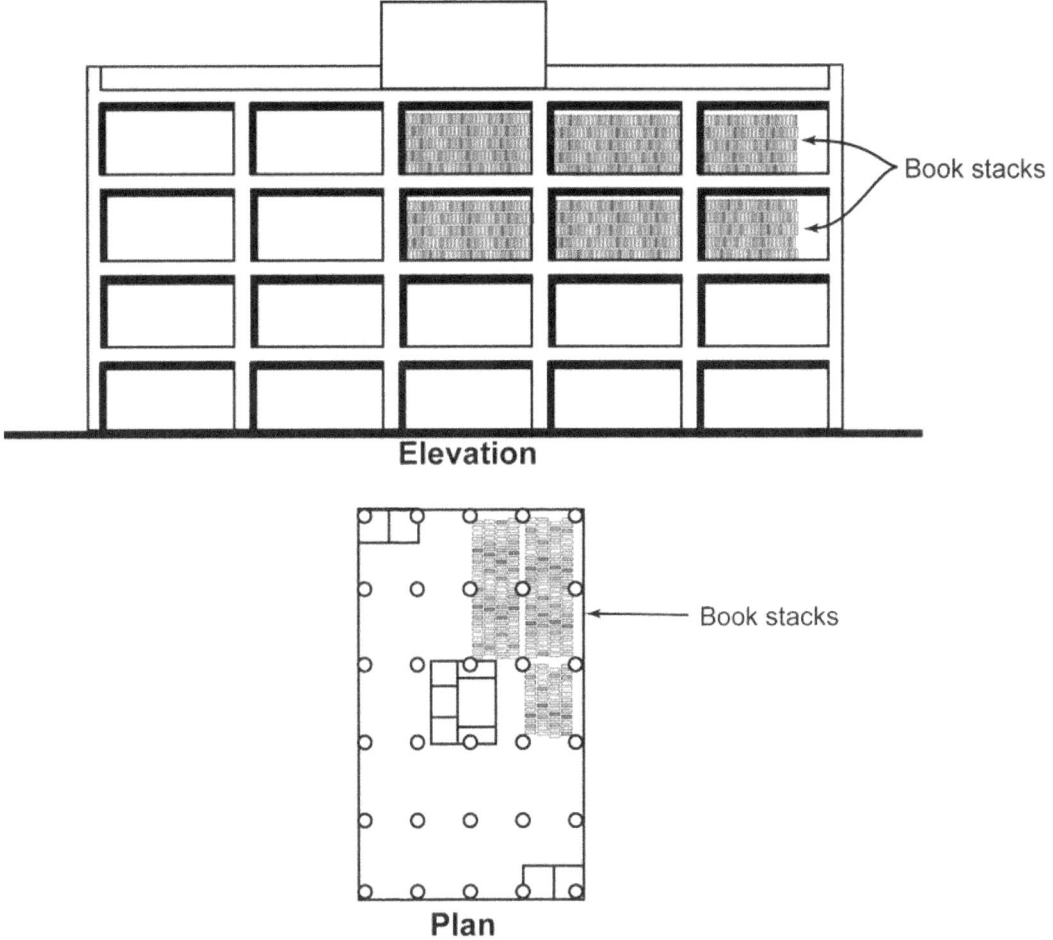

Elevation

Plan

← Book stacks

Figure 5-3   Nonsymmetric loading of book stacks in library building. Position and weight of stacks could induce torsional response of building during earthquake shaking.

○ In smaller buildings, stairs can act as bracing members between floors, introducing torsion; the solution is to detach the stair from the floor slab at one end to allow free structural movement.

○ In storage areas or library stacks, heavy storage items can introduce torsion into a structure. The structure may have been calculated to accommodate the maximum dead load but consideration be lacking for the effect of nonsymmetric loading over time as, for example, when library books are acquired (Figure 5-3).

5.5 QUANTIFYING THE BENEFITS OF IMPROVED PERFORMANCE

The benefits of improved performance are the reduced losses resulting from improved performance. These reduced losses include not only the reduction in capital losses (as described below), but also the reduction in financial impacts resulting from the loss of operations.

The benefits of improved earthquake performance of a building are quantified differently by the various types of building owners and users. For example, an owner occupant, an owner of a tenant-occupied building, and a tenant will all have different priorities and views regarding the cost-benefit trade-offs associated with improved earthquake performance of the building.

From the point of view of an owner occupant, earthquake performance of a building can be quantified in terms of reducing the probability of:

○ deaths and injuries in and around the building caused by an earthquake, and the resultant liability;

○ collapse of the building or damage to the building that reduces the building's value;

○ disruption of building services (HVAC, plumbing, electrical) and the resultant loss of use of the building or portions of it;

○ damage to building contents such as furniture, files, and inventory; and

○ disruption of building operation and business as a result of the above.

From the point of view of an owner of a tenant-occupied building, earthquake performance of a building can be quantified in terms of reducing the probability of:

○ deaths and injuries in and around the building caused by an earthquake, and the resultant liability;

○ collapse of the building or damage to the building that reduces the building's value; and

○ disruption of building services (HVAC, plumbing, electrical) and the resultant loss of use of the building or portions of it (tenant business interruption).

From the point of view of a tenant who is not the owner, earthquake performance of a building can be quantified in terms of reducing the probability of:

○ disruption of building services (e.g., HVAC, plumbing, electrical) and the resultant loss of use of the building or parts of it;

○ damage to building contents such as furniture, files, and contents; and

○ disruption of building operation and business as a result of the above.

Quantifying Expected Capital Losses

Capital losses consist of the cost of replacing or repairing earthquake-damaged structural and nonstructural compo-nents as well as damaged building contents. When quantify-ing capital value of damaged building components and contents, the first distinction that needs to be made is between the depreciated value of an asset, its market value, and its replacement cost. The assumption is generally that damaged capital will be replaced. If a 50-year-old building or piece of equipment is damaged to the extent that it is a total loss, it is unlikely that an owner can have a replacement building constructed or purchase a new piece of equipment for the same price as the original cost, nor for the depreciated value of the building or equipment (which may be very low, or zero). One may purchase a replacement building or piece of equipment that is 50 years old. Still, the price of that building or equip-ment will be based not on the depreciated value, but on the current market value of the asset. When the cost of losing an asset is evaluated, the owner must therefore determine what the cost to replace the asset will be, whether it is new or used.

An owner can use various means to estimate the replacement cost of a building or its contents. Realtors, manufacturers, engineers, and other specialists can research market conditions to estimate costs. If the owner has a large number of facilities or buildings, he/she may have a database of recent capital projects from which to draw information.

If structural and nonstructural building elements suffer less than total loss in an earthquake, they can often be repaired without being replaced. Theoretically, one would never spend more than the replacement cost of the building to repair structural and nonstructural damage. Practically, most owners con-sider the limit of repair costs to be on the order of 40 to 60 percent of the replacement cost. The older the building or equipment, typically the lower the threshold. The reason-ing is that if a building or piece of equipment is old and outdated, repairing it leaves the owner with something that, although functional, is still old and possibly outdated.

Owners may have other constraints which raise or lower this threshold. If short on cash or credit, an owner may have no choice but to repair a

Quantifying Capital Value

When quantifying capital value, the first distinction that needs to be made is between the depreciated value of an asset, its market value, and its replacement cost.

Repair Costs

Most owners consider the limit of repair costs to be on the order of 40 to 60 percent of the replacement cost. The older the building or equipment, typically the lower the threshold. Owners, however, may have other constraints which raise or lower this threshold.

heavily damaged building rather than replace it. If the operations within the building are so valuable that losses from down time far exceed the building's replacement cost, then even if very expensive repairs can be done more quickly than replacement, the threshold of repairable damage may also be high. On the other hand, if an owner has been looking to get rid of an old, poorly configured, structure even small amounts of damage may provide a convenient excuse to replace the building.

Repair Cost Vs Cost of New Construction

It should be clear that unit repair costs are rarely equal to the unit costs of new construction — they are typically higher.

It should be clear that unit repair costs are rarely equal to the unit costs of new construction. The cost of building partition walls in a new building, for example, may be on the order of five dollars per square foot. Repairing heavily damaged partition walls may cost more than twice this amount. Removing and replacing a damaged steel brace within a building may cost several times the cost of installing the brace in a new building.

A key to estimating the cost of repairing structural and nonstructural damage is understanding what the nature of the damage may look like. This is often defined as the "fragility" of the building system. Fragility presents the likelihood of damage as a function of the forces or deformations imposed on the building. Damage may be described in terms that include cracking or spalling of concrete elements; fracturing or buckling of steel beams, columns, or braces; glazing breakage; and partition cracking or failure. Estimating how much damage occurs at a specific stress or deformation has been and continues to be the subject of research. Once estimates of the damage are made, contractors and cost estimators can provide valuable assistance to owners and the design team in estimating repair costs.

Quantifying Replacement Costs

Damage to contents and inventory is usually quantified in terms of the amount of each that needs to be replaced.

Damage to contents and inventory is usually quantified in terms of the amount of each that needs to be replaced. In some cases, with very expensive equipment or inventory, one might consider repairing damage. In most instances, however, damaged items are typically replaced. Damage to non-fixed items typically occurs as a result of high accelerations "flinging" items off shelves or overturning them. Earthquake-induced accelerations vary over the height of a building so that items in upper stories may be more prone to damage than at lower stories. Estimating the amount of damage to contents and inventory involves calculating the acceleration at each level and estimating the capacity of elements at each story to withstand these accelerations. Shelving should be evaluated as to its overturning capacity and the potential damage of items that are spilled.

Contents such as desks and cabinets are fairly resilient to damage from sliding or falling, and are typically considered as losses only when they cannot be recovered because of substantial structural damage. Therefore, one might consider a threshold of structural damage (say when the building is condemned following an earthquake, or when it reaches its replacement threshold) at which point most of the contents are considered lost.

Quantifying Loss of Operations

Structural and nonstructural damage may require that a building's operations be curtailed or cease altogether for some period during repair or replacement. The loss of operations will have a direct effect on the revenue or "value" of the services or goods that the business produces. It will also, presumably, have a broader impact on its employees, on the customers that it serves, and possibly on the community or region as a whole. Business interruption may also be a factor in how soon, if ever, the business can recover lost opportunities and markets.

The primary impacts caused by loss of operations include:

○ Direct loss of revenue or value;

○ Indirect losses to employees, customers, and the community at large; and

○ Long term business losses.

All three of these impacts are dependent on how long and to what extent the business is out of operation. This is usually a function of structural and nonstructural damage, and may also be a function of contents loss. The impact of loss of operations on two facility types are demonstrated in the two example case studies described on the following page.

The loss of function of any single building is unlikely to cause devastating consequences to people in the affected region; nonetheless, these losses can be severe if the affected facilities are critical to community functions or the local or regional economy. Following are example situations where the loss of a critical facility can negatively affect the community as a whole or have far-reaching consequences:

○ In August, the only high school in a city is damaged to the point where it must be replaced. Where do the students go to school for the coming year or more while a new facility is designed and built?

Loss of Operations Case Study: Data Center

Situation.

A tilt-up building used as a data center suffers damage that causes a partial closure such that until cracks in several shear walls are repaired, access can only be allowed for up to four hours a day by no more than ten employees.

Because of the vibration-sensitive equipment contained in the building and the need for constant structural monitoring, the limitation on access essentially means that only 25% of the data center can be operated and maintained until the cracks are repaired.

Repairs take six months after which time the data center is fully functional.

Impacts Resulting from Loss of Operations.

The data center, which provides server space for clients, will lose 75 percent of its revenue initially. Suppose that, after three months, enough of the space is repaired so that 50 percent of the center's capability can be restored. The direct losses could be 75 percent of its revenue for the first three months and 50 percent of its revenue for three additional months. Indirectly, however, the data center company may have to either pay its employees salaries during that time, or temporarily or permanently lay them off. In the latter case, the company may have to pay the expense of rehiring employees once the facility is fully functional. The company may also have to pay damages to clients that lost data because of the loss of operations.

Long term, the data center company may permanently lose the customers it wasn't able to keep while repairs were in process. These customers presumably need server space following the earthquake and during the center's repair, and would look elsewhere for it. Once they find alternate space, they may be reluctant to switch back. The company, therefore, may lose market share for some time until it can recover lost clients or generate new ones.

Loss of Operations Case Study: University Laboratory Building

Situation.

A university laboratory building is badly damaged after an event with losses greater than 60 percent of the replacement cost. The building and laboratory equipment are twenty years old; the university therefore makes the decision to replace the building.

The laboratory is highly specialized, and researchers are unable to proceed with their experimental work for the three-year duration of the building replacement. The new building will, however, contain state-of-the-art facilities.

Impacts Resulting from Loss of Operations.

In the second example, the university would presumably lose the direct revenue in grant funding it received for the research conducted in the laboratory. Because the university is a non-profit organization, and a significant portion (say, 1/3) of the grant revenue pays university overhead costs, which include campus-wide expenses, loss of the laboratory would have consequences that reach far beyond the loss of the laboratory facility. Such loss revenue could cause campus-wide reductions in staffing and other goods and services, depending on the ratio of overhead revenue lost versus overhead revenue amounts from other sources.

Beyond the immediate revenue losses in this example are the additional potential impacts if students, faculty, and staff elect to leave for other institutions if they cannot continue to conduct the research of their choice at the university. This could have a long and lasting impact on the output and future funding for the university, and may hurt its future ability to attract researchers and students. These considerations are not easy to quantify in dollar terms; they should, however, play an important role in determining the willingness to invest in a better performing building.

○ The only county hospital with a trauma center is rendered non-functional during an event that causes dozens of life-threatening injuries within the community.

○ A pharmaceutical manufacturing plant that produces a popular drug for which the company owns the patent is destroyed in an earthquake. How will patients continue to get the drug?

○ An automotive parts manufacturer that provides "just-in-time" supplies to an automobile maker cannot function for three months. How will this affect the automobile company's ability to produce cars and its ability to keep its employees busy?

It is almost impossible to put a dollar value on the cost of these losses because, like many other events, the repercussions can be difficult to completely define. It is therefore unrealistic to develop a pure cost–benefit study equating additional dollars spent on better performance with savings in terms of these reduced indirect effects.

COST CONSIDERATION

It is almost impossible to put a dollar value on the cost of indirect losses. One can, however, make comparative studies with respect to other types of risks and establish an equivalent value of tolerating them.

One can, however, make comparative studies with respect to other types of risks and establish an equivalent value of tolerating them. In any of the examples above, the building owner will likely have liability insurance to protect against claims that could have a devastating impact on the entity. A private school might have a catastrophic insurance policy to protect against a student being killed in a sporting event; a public school may have locally- or state-granted legal protections. A hospital certainly has malpractice insurance and an automotive plant will have worker's compensation insurance. However, insurance policies all have limits on coverage. If losses exceed the coverage limit the result could be bankruptcy. Yet the owner in all cases makes a decision to limit coverage and therefore to accept the remaining risk of catastrophic loss.

Considering the example of the school injury, if a family of an injured student wins a judgment exceeding the school's insurance policy, the school may have to declare bankruptcy and close its doors. The school may be able to avert this consequence if it buys additional insurance. However, at some point it makes the decision that it is not going to spend more in premiums and is willing to accept the risk of a catastrophic loss. The process to arrive at this limit may have been explicitly or implicitly thought out. Regardless, it can be used as a guide for making other decisions about risk management for earthquakes. The case study icon (see next page) illustrates this hypothetical situation.

When determining the level of performance for which a building should be designed, an owner may want to consider involving those outside the business who will be indirectly affected by the potential loss of operations. A community might be willing to contribute to the cost of a higher performance design of a school if it considers the value of having the building usable after an earthquake sufficiently high. Similarly, an auto maker might contribute to the performance-based design of one of its parts suppliers if it considers an uninterrupted supply of parts crucial to its own operations.

CASE STUDY

Comparative Risk Tolerance Case Study: Seismic Risk Management Versus Student Injury Liability Insurance

Student injury liability policy:

○ Up to $1,000,000 per incident (excludes earthquakes)

○ Annual premium: $40,000

○ Out-of-pocket loss above which would result in school bankruptcy: $2,000,000

○ Total manageable loss: $1,000,000 (insurance) + $2,000,000 (out-of-pocket) = $3,000,000

○ Annual likelihood of a $3,000,000 claim: 1/2%

   Risk Tolerance: willing to spend $40,000 annually to limit risk to a 1/2% chance of a catastrophic loss.

Earthquake risk management situation:

○ School is planning to move to a new site and build new facilities.

○ Earthquake ground motions with a 1/2 % probability of being exceeded per year, (which correspond to a 200-year return period) are expected to cause $500,000 in capital losses, relocation for six months at a cost of $500,000, and injury to students at a cost of $1,500,000 = $2,500,000 total.

○ Risk of 1/2 % for a $2,500,000 loss exceeds threshold ($2,000,000) established for student injury liability (see above)

   Comparable Risk Tolerance: If the premium for earthquake insurance is no more than $40,000 to cover $500,000 of capital losses, then spending on earthquake risk reduction is at least as good an investment as the liability policy.

Social and Political Factors Affecting Seismic Risk Management

Emotion and politics are often important factors in the seismic risk management decision-making process. Parents of school children may say, "No price is too high to pay for the safety of my child." Politicians or business leaders may proclaim, "We have a zero tolerance policy for placing the occupants of our buildings at risk." While well intended, these positions are not often achieved in practice.

The concept of placing a quantified (dollar) value on the life or safety of each person is a controversial issue that impacts benefit-cost analysis. This approach is implemented by comparing the value of saved lives (the benefit) to the cost of protecting those lives. Political or emotional constraints often make this extremely difficult. If an owner looks beyond life safety, however, and focuses on capital losses or down time, then it is practical and possibly necessary to quantify these losses in terms that can be compared directly to the costs to reduce them. The fact that most new U.S.-code-designed buildings are expected to provide life safety (for the range of earthquakes that may occur over the life of the building) renders the need to assign dollar values to human lives less imperative.

5.6 COSTS OF IMPROVED PERFORMANCE

Building owners incur costs to obtain specified levels of building performance. These costs are considered "first costs" if incurred at the time of building design and construction or purchase. They are considered "operating costs" if incurred over the period of use of the building.

It should be noted that the period of use of a building by its owner might differ from the life of the building. The life of a building may be 60 years or more, while the owner's use could be much shorter. When considering societal costs of a building, for example energy use, society's interest in operating costs are spread over the life of the building, regardless of owner turnover. The life of the building is also of interest in the operating cost considerations of certain types of owners, particularly institutional owners such as schools and universities. However, for most commercial owners considering making an investment in a building, operating costs are of interest only over the period that the owner anticipates owning the building.

First Costs

The following are typical of first costs:

○ The costs of site selection, including the cost of physical and economic analysis of alternative sites.

○ The costs of planning and programming a new building, including the costs of consultants.

○ The costs of architectural and engineering design and construction management, in the case of the construction of a new building, or the transaction costs (e.g., inspection and appraisal), in the case of the acquisition of an existing building.

○ The disruption of operations resulting from the move from a currently used building to a new building.

Except for the last item, there is generally a direct relationship between cost and building performance (including seismic performance) – a higher first cost investment typically results in improved performance.

Operating cost analyses often categorize the costs of construction or purchase as first costs. This is short sighted in most cases, since these costs are usually financed through mortgages or bonds, which converts them into continuous operating costs.

Operating Costs

The following are typical operating costs:

○ Operation and maintenance of the building, including costs of earthquake response and recovery.

○ Replacement of building components and systems, including the cost of disruption of operation related to these activities, both of which can be annualized if converted to a payment into a replacement reserve fund.

○ Changing the building to accommodate new functions or technology, and the disruption of operation resulting from such activities (which is analogous to churn rate).

○ Insuring the building. Higher costs in this category may improve building performance by reducing unrecoverable losses or they may be inversely related to it, depending on insurance company underwriting practices.

○ Building and contents damage resulting from unpredictable events, such as natural and man-made disasters, which can be expressed as a probability of incurring an annual cost.

○ Disruption of operation due to building damage resulting from unpredictable events, which can be also expressed as a probability of incurring an annual cost.

○ Liability for deaths and injuries from building damage resulting from unpredictable events.

IMPROVING PERFORMANCE TO REDUCE SEISMIC RISK

An advantage of performance-based design is that it provides a means for the design team to create a relationship between construction cost and performance. Traditionally (i.e., using existing seismic codes for new building design), to achieve better performance an engineer might simply increase the importance (I) factor from 1.0 to 1.25, thereby raising the design seismic forces by 25%. This may make a building perform better; however, the benefit is not easily quantifiable, even if the cost in increased steel tonnage or concrete volume can be estimated.

A more refined way of achieving a specified performance in a cost efficient manner is to develop "learning curve" type relationships between the two. Consider the example in Figure 5-4. A hypothetical precast concrete tilt-up manufacturing facility is to be constructed in a moderately high seismic zone. The lowest cost for the building is that which meets the minimum requirements of the building code. At this design level, the building will be expected to suffer some loss in the "worst case" earthquake, however that is defined. Additional investments in

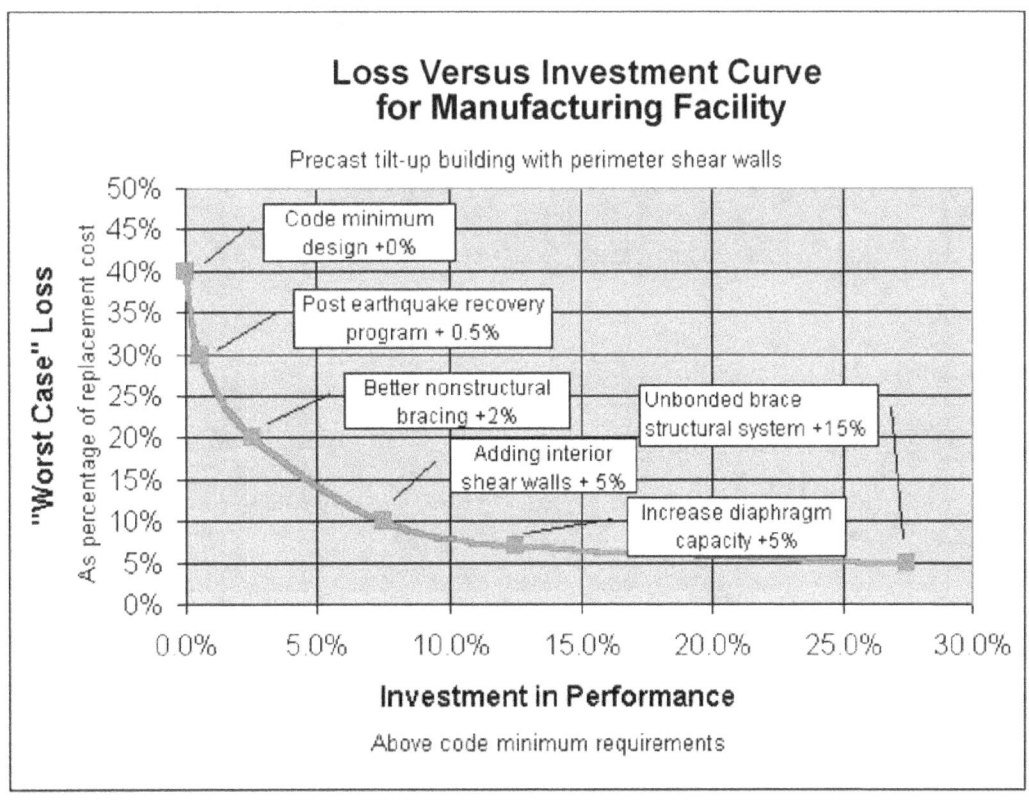

Figure 5-4   Relationship between cost and performance for hypothetical example.

---

improved performance might be considered by the design team and owner. If the cost for each investment is added cumulatively as each is included in the construction budget then the expected worst case loss should decrease. As this example shows, investments in postearthquake response and nonstructural bracing result in a relatively large benefit in terms of reduced losses. Adding interior shear walls results in a moderate benefit. Increasing diaphragm strength and changing the entire structural system to unbonded braces produce a relatively low benefit.

This example can be taken further by computing the benefit-cost ratio (BCR) for each performance strategy. Suppose the likelihood of the worst case event occurring over a 50-year life of the building is 25%, which corresponds to a 0.58% annual probability of occurrence[2], and that the code minimum cost is equal to the replacement cost. Assuming a 5% discount rate (rate of return), the resulting benefits and costs are as summarized in Table 5-5.

Table 5-5   Summary of Benefits and Costs for Hypothetical Manufacturing Facility Example

| Risk Reduction Strategy | (As Percent of replacement cost) | | | | BCR (Benefit/ Cumulative Cost of investment) |
	Cumulative Cost of investment (above Code Minimum)	"Worst Case" Loss	Present Value of Loss[1]	Benefit[2]	
Code minimum design	0.0%	40%	4.2%	0.0%	-
Post earthquake recovery program	0.5%	30%	3.2%	1.0%	2.1
Better nonstructural bracing	2.5%	20%	2.1%	2.1%	0.85
Adding interior shear walls	7.5%	10%	1.1%	3.2%	0.42
Increase diaphragm capacity	12.5%	7%	0.7%	3.5%	0.28
Unbonded brace structural system	27.5%	5%	0.5%	3.7%	0.13

Notes:

[1]Computed as: $PV = pmt \left[\frac{1}{r} \cdot \frac{1}{r(1+r)^n} \right]$ where PV = the Present Value of Loss; pmt = Annual Loss ("Worst Case" Loss times the annual probability of occurrence); r = rate of return; and n = life of building in years

[2]Present Value of Loss for code minimum design less Present Value of Loss with cumulative investment in performance

---

2. Computed as (ln(1-probability of occurrence in n years))/(-n years)

The example suggests that, in this case, the incorporation of a post-earthquake recovery program with a BCR of 2.1 is clearly a good investment. Improving nonstructural bracing in addition results in a BCR of 0.85 suggesting that it is possibly a good investment. The other performance strategies appear not to be economically beneficial.

A careful study of possible design strategies may lead to several cost–performance curves, such as Figure 5-4, incorporating different combinations of performance strategies. These will then allow the owner and the design team to select the one that achieves the greatest expected return on the investment.

5.7 CASE STUDIES OF COST AND PERFORMANCE CONSIDERATIONS

The following five case studies illustrate how different owners have addressed cost and performance considerations in seismic risk management decisions.

Case Study 1: Computer Graphics Equipment Maker in Salt Lake City, Utah

Salt Lake City, Utah is the headquarters of a small computer graphics equipment maker, as shown in Figure 5-5. The company's main products are high-end simulation systems that sell for nearly $10 million each. Its new corporate office was to include a large assembly floor in

Figure 5-5   Site of facilities for computer graphics equipment maker in Salt Lake City, Utah.

which eight to ten of these devices would be assembled at one time, as well as a floor of office space above. All of the company's manufacturing would be housed within the building. The loss of a single simulation device as a result of an earthquake would have caused a catastrophic loss for the company, resulting in possible bankruptcy.

The local structural engineer of record was skilled in performance-based design and well known in Salt Lake City because of his efforts to expand awareness of seismic issues. The engineer was able to develop a relationship with the owner directly, although he was part of a design team headed by an architect. This "access" to the owner was crucial in providing an opportunity for the engineer to explain concepts of performance-based seismic design. He and the owner discussed critical structural issues that could affect building performance and impact repair costs and business restoration.

The code in force at the time of construction would likely have protected the building against most earthquakes. The seismicity in the Salt Lake City region during a typical 30-50 year building life is relatively low. However, considering the consequence of damage and lost functionality, even the relatively low vulnerability still resulted in an intolerably high risk to the owner.

Because of the extremely high value of contents and cost of lost operations, a performance objective was established to limit structural and nonstructural damage in a rare event to a level that would protect the contents and allow operations to continue unimpeded.

To achieve this performance objective, the building was base isolated. The project team justified the additional cost associated with a base isolated building over a conventional structure by noting that the cost of the isolated structure was still less than the value of a single simulator. The vulnerability of the enhanced building was substantially lower than would be for a similar conventional structure. Much of the equipment, including the simulator devices, was braced to prevent tipping or sliding. The overall reduction in risk achieved was dramatic and met the owner's risk threshold. To reduce the risk any further, the building would likely have to have been re-sited to a region of lower seismicity.

Case Study 2: Salt Lake City, Salt Lake City K-12 School District

The Salt Lake City School District consists of 30-40 sites and contains buildings more than 70 years old, as shown in Figure 5-6. The District

Figure 5-6   Sample existing school building in Salt Lake City K-12 School District.

embarked upon a program of seismically upgrading its buildings to ensure that they would be safe and usable following a major seismic event. The District made the determination that it wanted to achieve a 70-year additional life for its structures.

Its study of the existing school facilities found that when nonstructural rehabilitation costs (e.g., heating, electrical, roofing, and deferred maintenance) were added to the structural costs necessary to achieve the high performance objective, many of the rehabilitations would cost more than the replacement cost of the building. In these cases the decision was made to replace the facilities with new designs such as that shown in Figure 5-7.

Figure 5-7   Sample new school building design in Salt Lake City K-12 School District.

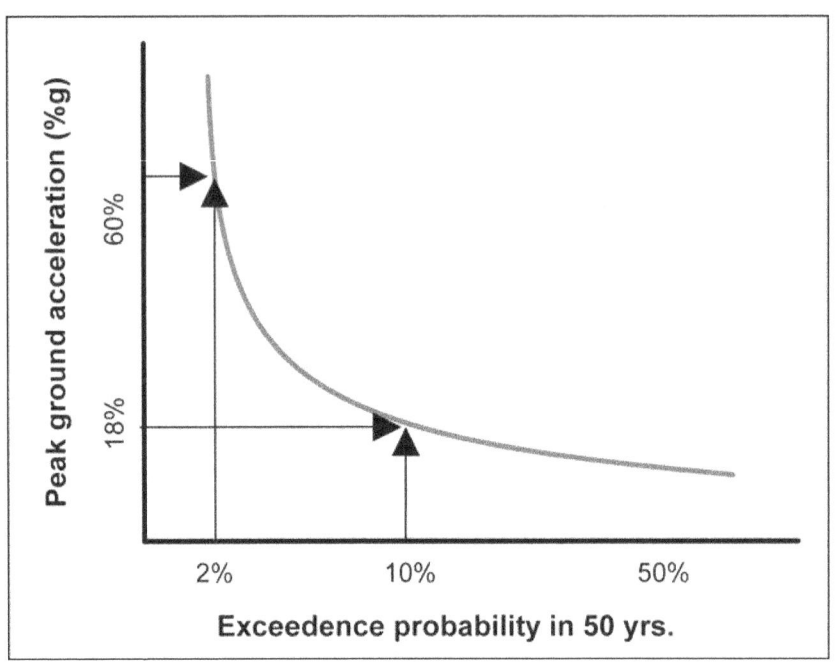

Figure 5-8   Site hazard curve for Salt Lake City K-12 School District seismic risk study.

Recognizing that building codes can change dramatically even over the course of ten to twenty years, the District asked its engineering consult-ant to evaluate the performance needs with its long lifetime in mind. The engineer crafted simple yet effective graphics similar to the plot shown in Figure 5-8 to help the District determine its risk tolerance. A site hazard curve (see discussion in Section 3.2) was developed (see Figure 5-8) to show the expected ground accelerations plotted against their probability of exceedence in 50 years. Salt Lake City is in *Uniform Building Code* (UBC) Seismic Zone 3, and the vertical line at 10% show the design ground motion specified by the UBC. Over a 70-year period, the probability of exceedence of this level of ground shaking increases from 10% to 14%. Another vertical line is drawn at 2% probability of exceedence in 50 years, representing perhaps the maximum credible event in the area. Over a period of 70 years, the probability of exceedence of this level of ground shaking increases from 2% to approximately 3%. Most notable is the dramatically higher ground motions that would be expected in the 2% probability of exceedence in 50-year event. This analysis showed that designing only for ground motions having a 10% probability of exceedence in 50 years meant there was still a risk of much higher ground motions that could seri-ously damage the facilities. The District wanted to achieve a higher level of confidence than 14% over the 70-year lifetime that damage would be kept to a minimum. The design forces for the 2% probability of exceedence in 50-year event compared well to the UBC Seismic Zone 4 design forces. Therefore, the District decided to design all its new facilities to Zone 4 requirements for forces and detailing, pending a cost analysis of the upgraded performance.

The consulting engineer estimated that to enhance a new facility's design from Zone 3 to Zone 4 compliance would add a cost on average of ¼ to 1% of the construction budget. The District quickly realized that amortized over the length of its construction financing and certainly over the length of the 70-year assumed lifetime, this additional cost was negligible and therefore adopted the enhanced design strategy.

Key factors in the owner's decision to use an enhanced performance objective were the expected longevity of the facilities and the number of buildings in the portfolio. The importance to the community of the school district, the large capital investment that was being made over the entire inventory, and the not inconsiderable likelihood of a damaging event occurring over the lives of the buildings were also important considerations in the District's decision.

Case Study 3: Prosthesis Manufacturing Company in Memphis, Tennessee

A prosthesis manufacturing company in Memphis, Tennessee was nearing completion of a 100,000 square foot manufacturing plant in early 2002. The products manufactured within the building generate revenues of nearly $500,000 per day. The building operations are insured against down time by a large international insurance company.

The building was built to the structural and nonstructural requirements of the 1997 *Southern Building Code* (SBCCI, 1997). The insurer offered to reduce the building's insurance premiums significantly if the nonstructural bracing was brought into conformance with the more severe requirements in the 2000 *International Building Code* (ICC, 2000). The *International Building Code* (IBC) requires that nonstructural bracing be designed to consider site conditions including soil and proximity to faults, and the location of the equipment within the building.

Cost-Effectiveness of Nonstructural Bracing

Improving the seismic nonstructural bracing in new buildings located in moderate seismic zones can be very cost efficient in terms of reducing losses

A New York based manufacturer and supplier of mechanical equipment bracing products was hired to assess the additional cost of bracing the equipment to the higher standard. Typically, the IBC design required only larger elements (e.g., anchor bolts, clevises, rods) and not a substantial change to the design configuration.

The manufacturer performed a detailed comparison of the two codes and prepared a side-by-side comparison of the cost premium for the IBC design. An excerpt from the comparison is shown in Table 5-6.

Table 5-6   Comparison of Costs for Design to SBCCI and IBC Requirements

EQUIPMENT	FOR SBCCI REQ.	FOR IBC REQ.	SBCCI PRICE	IBC PRICE
Fan Powered Boxes A Thru R	Cable Bracing (Spec 12)	Cable Bracing (Spec 12)	$92 Each. + $300 Total for calculation	$92 Each. + $300 Total for calculation
Fans F-1,2,3,4, In Line	Isolation Hangers, (Spec 11) Cable Bracing (Spec 12)	Isolation Hangers, (Spec 11) Cable Bracing (Spec 12)	$408+ $300 Total for calculation	$616+ $400 Total for calculation
F-5 Cabinet	Anchor Bolts (Spec 19), Grommets (Spec 14)	Anchor Bolts (Spec 19), Grommets (Spec 14)	$56+ $300 Total for calculation	$56+ $300 Total for calculation
F- 6, 7, 11, 12, 13, 14, 15, 17, 18, Rooftop	Mason Rigid Roofcurb	Mason Rigid Roofcurb	$43/Foot	$43/Foot
F-8, 9, 10, Wall Fans	Nothing Required	Nothing Required		

The results of the study showed that the additional cost of upgrading the seismic bracing was negligible as a percentage of the overall non-structural costs. The company decided, based on the benefit of reduced insurance premiums, to implement the higher standard.

This example suggests that improving the seismic nonstructural bracing in new buildings located in moderate seismic zones can be very cost efficient in terms of reducing losses. It may also result in the direct benefit of reduced annual insurance costs.

 CASE STUDY

Case Study 4: Stanley Hall, University of California Berkeley

The University of California (UC) at Berkeley is one of the nation's premier research institutions. In 2003 the university broke ground for a state-of-the-art bio-engineering laboratory building as shown in Figure 5-9. The estimated cost of the project is nearly $200 million. The building will contain high-end laboratory facilities and house researchers working annually on nearly $40 million in grants. UC Berkeley sits

Figure 5-9   New bioengineering laboratory building designed for UC
              Berkeley campus.

astride the active Hayward Fault. In the next thirty years, the USGS predicts there is nearly a 30% likelihood of a magnitude 7 or greater earthquake occurring on the fault (Working Group on California Earthquake Probabilities, 2003).

The University considered the protection of its massive investment in this facility to be extremely important. It asked the engineer of record to consider a higher performance objective than would have been required by the building code in force at the time. The goal was that the facility should remain occupiable after a design level event, and repair time to restore full operability should be measured in weeks not months. The engineer employed a state-of-the-art buckling restrained (unbonded brace) braced frame system to ensure that damage would be kept to a minimum even in a large event that might rupture the entire length of the Hayward Fault.

In order to obtain financing for the project, the University had to justify the added expense of the enhanced structural scheme. The school hired a second engineering firm expert in risk analysis, to help them provide the necessary rationale. The firm developed a "baseline" structural scheme that met only the minimum requirements of the building code. This system employed conventional concentric braced frames. The difference in cost between the two schemes was approximately $1.2 million, or roughly ½% of the building cost. They then used nonlinear performance-based engineering and risk assessment tools to calculate the expected losses due to earthquakes over the life of the building. The analysis showed that losses were substantially reduced using the enhanced scheme. The overall return on the $1.2 million investment,

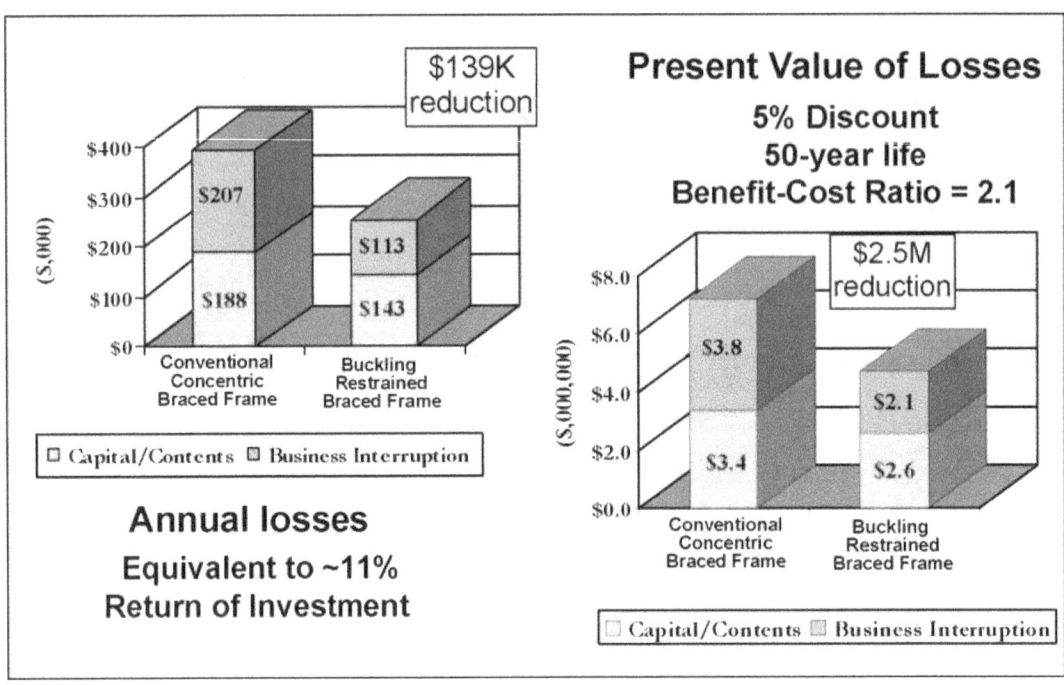

Figure 5-10   Comparison of future losses for two different structural system options for new UC Berkeley laboratory building.

considering reduction of capital, contents and business interruption losses was approximately 11%. Figure 5-10 shows that the reduction in business interruption provided the majority of the projected benefits. Using a 5% discount rate as a benchmark, the benefit-cost ratio (BCR) for the enhancements was over 2 considering a fifty year life. At that discount rate, the BCR reached one at a building life of about 15 years as shown in Figure 5-11.

This example suggests that performance-based design can be a very cost effective risk management strategy for buildings that generate substantial revenue and for which the owner has a long-term interest.

5.8   QUALITY CONTROL DURING THE CONSTRUCTION PROCESS

Quality control is an important aspect of assuring satisfactory seismic performance: the building must be constructed as designed and specified.

Building owners often interpret construction quality primarily in relation to interior and exterior finishes and materials because these are important for "marketing' in the private sector. It is generally assumed that design and construction to meet the applicable building codes will

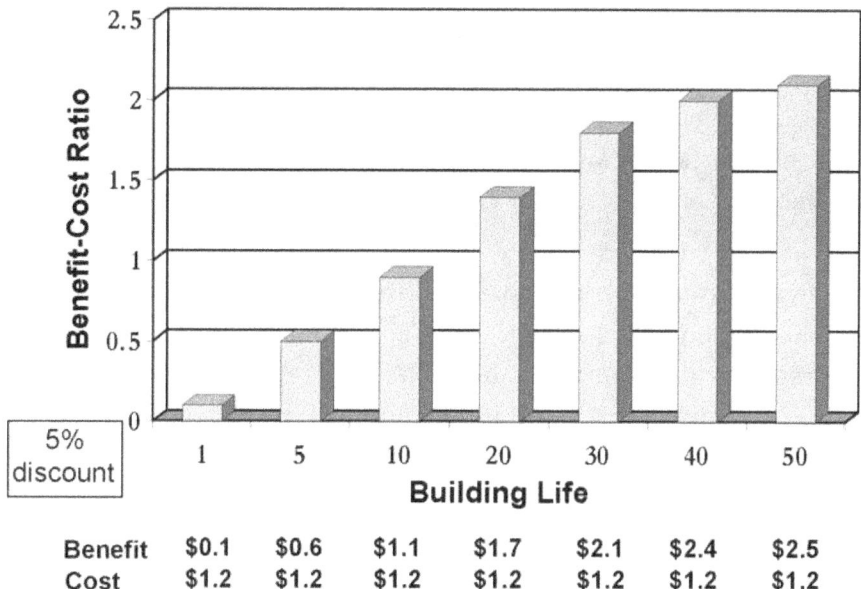

	1	5	10	20	30	40	50
Benefit	$0.1	$0.6	$1.1	$1.7	$2.1	$2.4	$2.5
Cost	$1.2	$1.2	$1.2	$1.2	$1.2	$1.2	$1.2

Figure 5-11   Comparison of benefit to cost of using a buckling retrained braced frame, instead of a conventional concentric braced frame, for new UC Berkeley laboratory building. Cost equals cost differential.

assure a durable and safe structure. Since structural elements are usually invisible—concealed behind a suspended ceiling, gypsum board or exterior cladding—they have little bearing on the perception of building quality. The exterior and interior appearance of the building will typically adhere to a company or institutional philosophy; this may be very functional for an industrial facility owner, but market trends or institutional objectives may influence others. The appearance of all facilities may also be influenced by local community design requirements. Decisions about image and quality have a major impact on construction cost, both initial and lifetime. See Section 12.5 for additional guidance on assuring design and construction quality.

5.9   RECOMMENDATIONS FOR IMPROVING SEISMIC PERFORMANCE

In addition to the specific seismic design issues relating to siting, structural systems, and nonstructural systems, there are some general measures that can be employed to help manage seismic risk by reducing either the vulnerability of the facility to earthquake damage, or the consequences of the damage should it occur. These measures include the following.

○ Consideration should be given to performance-based design to a level beyond Life Safety (typically the level of performance provided by provisions of the current seismic design codes) to a level of Immediate Occupancy, as discussed in Section 4.3. Institutional, public and corporate owners usually have long-term ownership of their facilities and a desire for continued operation in the post-event period.

○ The design professionals in charge of the structural and nonstructural component installation should specify quality assurance requirements; the contractor should be required to exercise a high degree of quality control; and independent inspection should be used to ensure conformance to requirements.

○ The design engineer should advise facility owners and manager on technical aspects of obtaining insurance to cover potential losses including service interruption. It may be possible to negotiate reduced premiums with the insurance carrier on the basis of any seismic mitigation measures provided beyond the code-minimum requirements.

○ Retainer agreements should be established with engineers and architects to provide building inspection services immediately following an earthquake (see Section 2.6 for additional information).

○ Personal protection and evacuation plans should be developed for all staff and students. Regular drills and educational sessions should be conducted to ensure proper execution (see Section 2.6 for additional information).

6.1 INTRODUCTION

Commercial office buildings represent a large building segment and house the core of American business operations. Corporate headquarters, banks, law firms, consulting firms, accountants, insurance companies, non-profit organizations – the list is almost endless – use office space in buildings around the country to house their operations. As these companies make decisions about the buildings that they construct or office space that they lease, seismic considerations can easily be factored into the decision process.

The following are some unique issues associated with commercial office buildings that should be kept in mind during the design and construction phase of new facilities:

- Protection of building occupants is a very high priority.

- Occupants are predominantly work-force, with high daytime "8 am to 5 pm" occupancy.

- Most office building occupants are generally familiar with the characteristics of their building; a small percentage of occupants may be disabled to some degree and visitors will generally not be familiar with the building.

○ Office buildings change their interior layouts frequently, to respond to tenant needs, fluctuations in work-force or organizational changes.

○ Ensuring the survival of business records, whether in electronic or written form, is essential for continued business operation.

○ Closure of the building for any length of time represents a serious business problem.

6.2 OWNERSHIP, FINANCING, AND PROCUREMENT

Commercial buildings may be owner operated, particularly if owned by national or global corporations, but many are developer owned (at least initially) housing tenant (lease holder) operations. In many instances the developer and building designers provide an empty "shell," which is fitted out according to the tenants' planning, spatial and environmental needs; design and construction is generally undertaken by the tenant's consultants and contractors. This tends to split the responsibility for interior nonstructural and other risk reduction design and construction measures between the building designers and contractor, and a multiplicity of tenant designers and contractors.

Financing for these facilities is typically through private loans. The effective life of an office building is 20 to 30 years, after which major renovation and updating is normally necessary. Interior renovation is usually on a much shorter interval, particularly for rental office structures.

6.3 PERFORMANCE OF OFFICE BUILDINGS IN PAST EARTHQUAKES

The seismic performance of modern office buildings designed to recent codes (adopted since the late 1970s) has been good as far as providing life safety. However, the recognition by building owners that satisfactory life-safety code-level performance may still encompass considerable damage (see Figure 6-1), along with repair costs and possible business interruption of the building for weeks or even months, even in a moderate earthquake, suggests that some performance-based design strategies may be useful.

Figure 6-1   Typical earthquake damage to contents and nonstructural components in a modern office building. (photo courtesy of the Earthquake Engineering Research Institute)

Where severe structural damage has occurred in commercial office buildings, it has generally been to older buildings, often the result of configuration irregularities. Figure 6-2 shows an older medical office building, which had a vertical irregularity that caused one floor to pan-cake during the 1994 Northridge earthquake in Southern California; a failure resulting from inadequate attachment of heavy nonstructural walls in an older 5-story office building is shown in Figure 6-3.

Newer office buildings have also been damaged, most notably the more than 100 welded steel moment-frame buildings (healthcare and resi-dential structures as well as commercial, higher education and indus-trial buildings) that failed during the 1994 Northridge earthquake. The damage occurred primarily at welded beam-to-column connections, which had been designed to act in a ductile manner and to be capable of withstanding repeated cycles of large inelastic deformation.

 While no casualties or collapses occurred as a result of these failures, the incidence of damage was sufficiently high in regions of strong motion to cause wide-spread concern by structural engineers and build-ing officials. Initial investigations showed that in some cases, 50% of the connections were broken and very occasionally the beam or column was totally fractured. Possible causes focused on incorrect connection

Figure 6-2   Exterior view of medical office building severely damaged by the 1994 Northridge earthquake. (C. Arnold photo)

Figure 6-3   Partially collapsed end-wall in 5-story office building caused by severe earthquake ground shaking. (C. Arnold photo)

design, incorrect fabrication, poor welding techniques and materials, and the impact of the need for economy on design strategies and construction techniques.

As a result, a large research program was initiated, sponsored primarily by FEMA, to identify the problems and arrive at solutions. Many structural specimens were tested in university laboratories. New guidelines for these types of structures have been developed (SAC, 2000a, b), but remedial measures have resulted in more costly designs and extended approval procedures, with the result that many engineers have avoided welded steel moment-resistant frames in recent projects.

Resources for the Seismic Design of New Steel Moment-Frame Buildings

1. FEMA 350, *Recommended Seismic Design Criteria for New Steel Moment-Frame Buildings* (SAC, 2000a)
2. FEMA 353, *Recommended Specifications and Quality Assurance Guidelines for Steel Moment-Frame Construction for Seismic Applications* (SAC, 2000b)

6.4 PERFORMANCE EXPECTATIONS AND REQUIREMENTS

The following guidelines are suggested as seismic performance objectives for commercial office buildings:

- Persons within and immediately outside the building must be protected to at least a life safety performance level during design-level earthquake ground motions.

- Persons should be able to evacuate the building quickly and safely after the occurrence of design-level earthquake ground motions.

- Emergency systems in the facility should remain operational after design-level earthquake ground motions.

- Emergency workers should be able to enter the building immediately after the occurrence of design-level earthquake ground motions, encountering minimum interference and danger.

6.5 SEISMIC DESIGN ISSUES

The information in this section summarizes the characteristics of commercial office buildings, notes their relationship to achieving good seismic performance, and suggests seismic risk management solutions that should be considered.

Seismic Hazard and Site Issues

Unusual site conditions, such as a near-source location, poor soil characteristics, or other seismic hazards, may lead to lower performance than expected by the code design. If any of these other suspected conditions are geologic hazards, a geotechnical engineering consultant should conduct a site-specific study. If defects are encountered, an alternative site should be considered (if possible), or appropriate soil stabilization, foundation and structural design approaches should be employed to reduce consequences of ground motion beyond code design values, or costly damage caused by geologic or other seismic hazards (see Chapter 3 for additional information). If possible, avoid sites that lack redundant access and are vulnerable to bridge or highway closure.

Structural System Issues

Office buildings are typically low- to mid-rise in suburban locations and occasionally high-rise in downtown locations of larger cities or in satellite suburban office complexes. Office buildings are intrinsically simple, and often are of simple rectangular configuration, not least because economy is usually a prime concern for commercial structures. Thus, their seismic design can be economical and use simple equivalent lateral force analysis procedures with a good probability of meeting code performance expectations as far as life safety is concerned. The protection of nonstructural components, systems and concepts requires structural design to a higher performance level. Configuration irregularities may be introduced for image reasons or site constraints in odd-shaped urban lots, and the structural design may become more complex and expensive. To assist the protection of nonstructural components, special attention should be paid to drift control.

The need for planning flexibility requires minimization of fixed interior structural elements and a preference for column-free space. Need for flexibility in power and electronic servicing has resulted in increasing use of under floor servicing to work cubicles, and structural systems have been developed to provide this.

Office buildings typically employ steel or reinforced concrete frames to permit maximum planning flexibility. Steel or reinforced concrete moment frames provide maximum flexibility, but tend to be expensive in high and moderate seismic zones. New guidelines for the design of welded moment-frame connections, noted above, have increased the cost of these types of structural system, increasing the already common use of steel braced frames. Elevator cores duct shafts and toilet rooms, being permanent, can be used as shear walls if of suitable size and location. Since these elements are much stiffer than a surrounding frame they may be a source of stress concentration and torsion, if asymmetrically located. If severe asymmetry of core locations is essential for planning reasons, the cores should not form part of the lateral-force resisting system.

Nonstructural System Issues

The extensive use of frame structures for commercial office buildings, together with the tendency for them to be designed to minimum code standards, has resulted in structures that are subject to considerable drift and motion (sway). The result has been a high level of nonstructural damage, particularly to partitions, ceilings and lighting. This kind of damage is costly and its repair is disruptive.

In addition, storage units, free standing work stations and filing cabinets are subject to upset. Excessive drift and motion may also lead to damage to roof-top equipment, and localized damage to water systems and fire suppression piping and sprinklers; thus the likelihood of water damage is greater.

The responsibilities within the design team for nonstructural component support and bracing design should be explicit and clear. The checklist for responsibility of nonstructural design in Chapter 12 (see Figure 12-5) provides a guide to establishing responsibilities for the design, installation, review and observation of all nonstructural components and systems

DESIGN AND PERFORMANCE ISSUES RELATING TO COMMERCIAL OFFICE BUILDINGS

DESIGN AND PERFORMANCE ISSUES RELATING TO RETAIL COMMERCIAL FACILITIES   7

7.1 INTRODUCTION

Retail commercial facilities house shops and stores, which contribute a signficiant portion of the nation's economic output. Department store malls, big-box retailers, grocery stores and strip malls are but a few of the almost endless list of retail operations housed in these types of facilities. As these companies make decisions about the buildings that they construct or spaces that they lease, seismic considerations can easily be factored into the decision process.

The following are some unique issues associated with retail commercial buildings that should be kept in mind during the design and construction phase of new facilities:

- Protection of building occupants is a very high priority.
- Occupants are predominantly work-force and shoppers; shopping malls and large retail stores typically are open from about 10 am to 9

pm for 7 days a week, typically with higher occupancy at weekends. "Big box" stores also have a high evening occupancy.

○ Most shoppers are generally familiar with the characteristics of the shopping malls stores they frequent, but large retail stores are confusing to the first-time shopper. Familiarity with exit locations and egress routes is questionable.

○ Retail stores, particularly department stores, change their interior layouts frequently to respond to market changes and retailing fashions. Big box stores generally retain a simple aisle layout, though some large electronic and furniture stores employ subdivided and clustered layouts related to groups of merchandise.

○ Ensuring the survival of business records, whether in electronic or written form, is essential for continued business operation.

7.2 OWNERSHIP, FINANCING, AND PROCUREMENT

Retail malls are generally developer sponsored. Department stores and "big boxes" are developed by regional or national owners; their design and construction are independent of the retail mall developments in which they may be located. In retail malls, the mall developer designs and constructs "shell" structures in which space is leased to retail store owners who use their own design and subcontracting teams to fit out the space to their requirements. This tends to split the responsibility for interior nonstructural and other risk-reduction design and construction measures between the building designers and contractor, and a multitude of tenant store designers and contractors.

Financing for these facilities is typically through private loans. The effective life of a retail mall or store is about 20 years, after which major renovation and updating is necessary. Interior renovation is usually on a much shorter interval.

Shopping malls and stores are generally constructed using a single contractor selected by competitive bid. Large shopping malls may have a number of contractors working on the site because each department store will usually have its own general contractor and subcontractors. Low cost and very rapid construction with reliable achievement of construction schedules are prime considerations. The opening of new retail facilities is often timed to meet key shopping periods such as Christmas or opening of the school year.

7.3 PERFORMANCE OF COMMERCIAL RETAIL FACILITIES IN PAST EARTHQUAKES

There has been considerable damage to retail facilities of all sizes in recent earthquakes.

In the Northridge earthquake of 1994 near Los Angeles, a large regional shopping mall with 1.5 million sq.ft. of retail space suffered severe damage and was closed for 18 months. Some 200 mall stores were closed and six department stores under independent ownership received varying amounts of damage. One department store suffered a partial collapse, and was demolished and replaced (Figure 7-1). The

Figure 7-1   Severe damage to a department store severely shaken by the 1994 Northridge earthquake. Shear failure between the waffle slabs and columns caused the collapse of several floors. (photo courtesy of the Earthquake Engineering Research Institute.)

other stores were repaired. Other shopping malls in the area suffered damage, but their performance was considerably better. The Topanga Plaza Mall in Canoga Park, approximately 5 miles from the epicenter, was built in the early 60's but was seismically upgraded in 1971. Structural damage was confined to cracking of reinforced masonry shear walls and damage to concrete columns in infilled shear walls. Nonstructural damage was significant, however, ranging from damage to floor, ceiling and wall finishes to frequently shattered or dislodged store-front glass panels.

7.4 PERFORMANCE EXPECTATIONS AND REQUIREMENTS

The following guidelines are suggested as seismic performance objectives for retail facilities:

- Staff and shoppers within and immediately outside retail stores must be protected to at least a life-safety performance level during design-level earthquake ground motions.

- Emergency systems in the facility should remain operational after the occurrence of design-level earthquake ground motions.

- Shoppers and staff should be able to evacuate the building quickly and safely after the occurrence of design-level earthquake ground motions.

- Emergency workers should be able to enter the building immediately after the occurrence of design-level earthquake ground motions, encountering minimum interference and danger.

7.5 SEISMIC DESIGN ISSUES

The information in this section summarizes the characteristics of retail facilities, notes their relationship to achieving good seismic performance, and suggests seismic risk management solutions that should be considered.

Seismic Hazard and Site Issues

Unusual site conditions, such as a near-source location, poor soil characteristics, or other seismic hazards, may lead to lower performance than expected by the code design.

Unusual site conditions, such as a near-source location, poor soil characteristics, or other seismic hazards, may lead to lower performance than expected by the code design. If any of these other suspected conditions are geologic hazards, a geotechnical engineering consultant should conduct a site-specific study. If defects are encountered, an alternative site should be considered (if possible) or appropriate soil stabilization, foundation and structural design approaches should be employed to reduce consequences of ground motion beyond code design values, or costly damage caused by geologic or other seismic hazards (see Chapter 3 for additional information). If possible, avoid sites that lack redundant access and are vulnerable to bridge or highway closure.

Structural System Issues

Retail facilities are usually one or two stories; mall structures and "big boxes" are usually light steel frames or mixed steel frame/wood/concrete/concrete masonry structures. Reinforced concrete block masonry perimeter walls often provide lateral resistance; for these systems, connections of roof diaphragms to walls are critical. The large building size and long-span light-frame load bearing structures of many of these facilities often lead to large drifts (or sway) during earthquake shaking. When designed to code minimums these drifts may be excessive and cause nonstructural damage, particularly to ceilings and partitions.

Retail buildings are intrinsically simple in their architectural/structural configuration, and basically are large open box-like structures with few interior walls and partitions. This enables their structural design to be simple and their seismic design can be carried out using the basic equivalent lateral force analysis procedures with a good probability of meeting code performance expectations as far as life safety is concerned. The desire for low cost, however, coupled with a tendency to meet only the minimum code requirements, sometimes results in inadequately engineered and poorly constructed structures. The protection of nonstructural components, systems and contents requires structural design to a higher performance level. Configuration irregularities are sometimes introduced for image reasons and the structural design may become more complex and expensive.

Nonstructural System Issues

The extensive use of light-steel-frame structures for retail facilities, together with the tendency for them to be designed to minimum codes and standards, has resulted in structures that are subject to considerable drift and motion. The result has been a high level of nonstructural damage, particularly to ceilings and lighting. This kind of damage is costly and its repair is disruptive.

In most "big box" stores the building structure forms only a weatherproof cover and is lightly loaded. Often there is no suspended ceiling and light fixtures are hung directly from the building's structure. The merchandise is stacked on metal storage racks, which provide vertical and lateral support. These racks are supplied and installed by specialist vendors. The correct sizing and bracing of these racks is critical because the merchandise is often heavy and located at a high elevation. Even if the racks remain, material may be displaced and fall on the aisles, which are often crowded.

More upscale department stores have complete suspended ceilings and often have elaborate settings for the display of merchandise. These can be hazardous to staff and shoppers.

Excessive drift and motion (building sway) may also lead to damage to roof-top equipment and localized damage to water systems and fire suppression piping and sprinklers.

The responsibilities within the design team for nonstructural component support and bracing design should be explicit and clear. The checklist for responsibility of nonstructural design in Chapter 12 (see Figure 12-5) provides a guide to establishing responsibilities for the design, installation, review and observation of all nonstructural components and systems.

DESIGN AND PERFORMANCE ISSUES RELATING TO LIGHT MANUFACTURING FACILITIES 8

8.1 INTRODUCTION

This chapter addresses a broad range of facilities used for industries engaged in the manufacturing assembly, testing and packaging of specialized products within workbench production areas. Much of this manufacturing is associated with the electronics, or "high-tech" industry, and in some cases, special environments such as "clean-rooms" are required. Most light manufacturing operations are relatively new and take place in recently designed and constructed buildings using modern equipment installations.

The following are some unique issues associated with light manufacturing facilities that should be kept in mind during the design and construction phase of new facilities:

○ Protection of building occupants is a very high priority.

- Building occupancy is relatively low, except in buildings with major production or assembly functions. Occupants are predominantly work-force, with high daytime "8 am to 5 pm" occupancy, although favorable market conditions may entail the use of additional work-shifts. Visitors are few in number.

- Ensuring the survival of production, testing and other expensive equipment is an important economic concern.

- Closure of the building for any length of time represents a very serious business problem, which will involve loss of revenue and possibly loss of market share.

- Most manufacturing building occupants are generally familiar with the characteristics of their building; a small percentage may be disabled to some degree.

- Frequent provision must be made for the production of new products and the removal of existing equipment and its replacement.

- Ensuring the survival of business records, whether in electronic or written form, is essential for continued business operation.

8.2 OWNERSHIP, FINANCING AND PROCUREMENT

Many light manufacturing facilities are owner developed, particularly if owned by national or global corporations, but some are also developer owned providing for tenant operations. Some large corporations may use a developer to build facilities that suit their operations, and thus avoid becoming involved in possibly troublesome development and building operations. Buildings that are constructed by developers as speculation tend to be occupied by start-up or young companies. In these instances the developer and building designers provide an empty "shell," which is fitted out according to the tenants' planning, spatial and environmental needs; design and construction is generally undertaken by the tenant's designers and subcontractors. This tends to split the responsibility for interior nonstructural and other risk-reduction design and construction measures between the building designers and contractor, and the tenant designers and contractors.

Financing for these facilities is typically through private loans. The effective life of the building may be about 50 years, particularly in the electronic industry. Light manufacturing buildings are generally constructed using a single contractor selected by competitive bid. Low cost and very rapid construction, with reliable achievement of construction schedules, are prime considerations.

8.3 PERFORMANCE OF LIGHT MANUFACTURING FACILITIES IN PAST EARTHQUAKES

Starting in the late 1950s larger light manufacturing buildings have been predominantly tilt-up structures, particularly in California. In seismic regions the perimeter precast walls were used as shear walls and roof structures were generally glued-laminated beams and plywood diaphragms. In the 1964 Alaska earthquake and the 1971 San Fernando (Los Angeles) event, performance of these buildings was poor, with considerable damage being sustained. The most common type of failure was to the wall/diaphragm anchors, but large out-of-plane movement of the panels, out-of-plane bending cracks in pilasters at mezzanine levels, and roof separations were all encountered and many roof collapses occurred. Due to the relatively large size of these buildings roof collapses were localized, rarely extending beyond one or two bays, and the buildings were sparsely occupied, so casualties were few. (Figure 8-1)

Following the 1971 San Fernando earthquake code changes were introduced, with the result that subsequent performance was improved. During the 1994 Northridge earthquake near Los Angeles, there were a number of failures of tilt-up structures and there were some collapsed wall panels along the sides of buildings resulting in partial roof collapse.

Figure 8-1   Roof and wall collapse of tilt-up building during the 1994 Northridge earthquake. (Photo courtesy of the Earthquake Engineering Research Institute)

Changes to wall anchorage requirements were introduced in the 1997 *Uniform Building Code.*

8.4 PERFORMANCE EXPECTATIONS AND REQUIREMENTS

The following guidelines are suggested as seismic performance objectives for light manufacturing facilities:

- ○ Persons within and immediately outside manufacturing facilities must be protected at least to a life-safety performance level during design-level earthquake ground motions.

- ○ Building occupants should be able to evacuate the building quickly and safely after the occurrence of design-level earthquake ground motions.

- ○ Emergency systems in the facility should remain operational after the occurrence of design-level earthquake ground motions.

- ○ Emergency workers should be able to enter the building immediately after the occurrence of design-level earthquake ground motions, encountering minimum interference and danger.

- ○ Key manufacturing equipment, supplies and products should be protected from damage.

- ○ In "high-tech" manufacturing facilities most services and utilities should be available within three hours of the occurrence of design-level earthquake ground motions.

- ○ There should be no release of hazardous substances as a result of the occurrence of design-level earthquake ground motions.

8.5 SEISMIC DESIGN ISSUES

The information in this section summarizes the characteristics of light manufacturing facilities, notes their relationship to achieving good seismic performance, and suggests seismic risk management solutions that should be considered.

Seismic Hazard and Site Issues

Unusual site conditions, such as a near-source location, poor soil characteristics, or other seismic hazards, may lead to lower performance than expected by the code design.

Unusual site conditions, such as a near-source location, poor soil characteristics, or other seismic hazards, may lead to lower performance than expected by the code design. If any of these other suspected conditions are geological hazards, a geotechnical engineering consultant should conduct a site-specific study. If defects are encountered, an alternative site should be considered (if possible) or appropriate soil stabilization, foundation and structural design approaches should be employed to reduce consequences of ground motion beyond code design values, or costly damage caused by geologic or other seismic hazards (see Chapter 3 for additional information). If possible, avoid sites that lack redundant access and are vulnerable to bridge or highway closure.

Structural System Issues

Light manufacturing facilities are usually one story; sometimes office/ administrative accommodation is provided in a mezzanine space. There has been increasing use of light steel frames and steel deck structure for roofs and mezzanines. Most large buildings now use braced steel frame structures. Exteriors may be of masonry or metal insulated panels.

Manufacturing buildings are intrinsically simple in their architectural/ structural configuration, and basically are large open box-like structures with few interior walls and partitions. This enables their structural design to be simple, and their seismic design can be carried out using the basic equivalent lateral force analysis procedures with a good probability of meeting code performance expectations as far as life safety is concerned. The desire for low cost, however, coupled with a tendency to meet only the minimum code requirements sometimes results in inadequately engineered and poorly constructed structures, The protection of valuable equipment and contents requires structural design to a higher performance level.

The large building size and long-span light frame load bearing structures of many of these facilities often lead to large drifts (or sway). When designed to code minimums these drifts may be excessive and cause nonstructural damage, particularly to ceilings and partitions.

Nonstructural System Issues

Continued operation is particularly dependent on nonstructural components and systems, including purchased equipment, much of which is often of great sensitivity and cost. Many specialized utilities must be provided, some of which

> **RISK CONSIDERATION**
>
> Continued operation is particularly dependent on nonstructural components and systems

involve the storage of hazardous substances, such as pharmaceuticals, or hazardous gases. These must be protected against spillage during an earthquake. Distribution systems for hazardous gases must be well supported and braced.

The extensive use of light-steel-frame structures for manufacturing facilities, together with the tendency for them to be designed to minimum codes and standards, has resulted in structures that are subject to considerable drift and motion. As a result, recent earthquakes have caused a high level of nonstructural damage, particularly to ceilings and lighting. This kind of damage is costly and its repair is disruptive.

Research and production areas may need special design attention to specialized equipment services and materials to ensure continued production and delivery.

In most manufacturing facilities the building structure forms only a weatherproof cover and is lightly loaded. Often there is no suspended ceiling and light fixtures are hung directly from the building's structure. In storage areas, materials are stacked on metal storage racks that provide their own vertical and lateral support. These racks are supplied and installed by specialist vendors. The correct sizing and bracing of these racks are critical if the materials are heavy and located at a high elevation. Even if the racks remain stable, material may be displaced and fall on the aisles or on equipment

Storage units, free standing work stations, and filing cabinets are also subject to upset. Excessive drift and motion may lead to damage to rooftop equipment and localized damage to water systems and fire suppression piping and sprinklers.

The responsibilities within the design team for nonstructural component support and bracing design should be explicit and clear. The checklist for responsibility of nonstructural design in Chapter 12 (see Figure 12-5) provides a guide to establishing responsibilities for the design, installation, review and observation of all nonstructural components and systems.

DESIGN AND PERFORMANCE ISSUES RELATING TO HEALTHCARE FACILITIES 9

9.1   INTRODUCTION

Healthcare facilities are the places where America goes for treatment for most of its healthcare and are the places that need to be available to them after being injured in an earthquake. Regional or local hospitals, outpatient clinics, long-term care facilities are all examples of healthcare facilities that serve in this role. As healthcare companies make decisions about the buildings that they construct, seismic considerations can easily be factored into the decision process.

The following are some unique issues associated with healthcare facilities that should be kept in mind during the design and construction phase of new facilities:

○ Protection of patients and healthcare staff is a very high priority.

○ Healthcare occupancy is a 24 hour/7 day-per-week function.

○ Acute-care hospitals have a large patient population that is immobile and helpless, for whom a safe environment is essential. This particularly requires a safe structure and prevention of falling objects.

○ Hospitals are critical for emergency treatment of earthquake victims and recovery efforts.

○ Medical staff has a crucial role to play in the immediate emergency and during the recovery period.

○ Ensuring the survival of all equipment and supplies used for emergency diagnosis and treatment is essential for patient care.

○ Ensuring the survival of medical and other records, whether in electronic or written form, is essential for continued patient care.

○ Closure of hospitals for any length of time represents a very serious community problem exacerbated by the possibility of the loss of healthcare personnel who are in high demand or unable to work because of personal earthquake-related consequences (e.g., their own injury).

○ Many hospitals are not only service providers but also profit or non-profit businesses and, since their operating costs and revenues are high, every day that the facility is out of operation represents serious financial loss.

9.2 OWNERSHIP, FINANCING, AND PROCUREMENT

Healthcare facilities are typically developed by a private non-profit or for-profit hospital corporation or an HMO (health maintenance organization). Many are also developed by a local, state or federal government agency. Financing of privately owned facilities is typically by private loan, possibly with some state or federal assistance; for-profit hospitals may issue stock when access to capital is required, and hospitals also conduct fund-raising activities, a large part of which assist in capital improvement program financing. State and local public institutions are financed by state and local bond issues. Non-profit hospitals sometime issue bonds to the public.

Private institutions have no restrictions on methods of procurement; projects may be negotiated, conventionally bid, use construction management or design–build. Public work must be competitively bid. Typically, contracts are placed for all site and building work (structural and nonstructural). Medical equipment and furnishings and their installation are purchased separately from specialized vendors.

Hospitals typically emphasize high quality of design and construction and long facility life, though all institutions are also budgeting conscious. An attractive and well equipped hospital site and building campus are seen as an important asset, particularly by private institutions that are in a competitive situation.

9.3 PERFORMANCE OF HEALTHCARE FACILITIES IN PAST EARTHQUAKES

The most significant experience of seismic performance of healthcare facilities in recent earthquakes was that of the Northridge (Los Angeles), California, earthquake of 1994. The San Fernando, California, earthquake of 1971 seriously damaged several medical facilities, including the then brand-new Los Angeles County Olive View Hospital. Most of the fatalities in this earthquake occurred in hospitals, principally the result of the collapse of an older unreinforced masonry Veterans Hospital building. In response to the recognized need for superior seismic performance by hospitals, the California Legislature enacted the Alfred E. Alquist Hospital Facilities Seismic Safety Act, which became effective in 1973. This Act mandated enhanced levels of design and construction. The Act proved very effective in limiting structural damage in the Northridge earthquake; no post–Act hospitals were red-tagged (posted with a red UNSAFE postearthquake safety inspection placard) and only one was yellow-tagged (posted with a yellow RESTRICTED USE placard). However, nonstructural damage was extensive, resulting in the temporary closure of several of the post-1973 buildings and the evacuation of patients.

Long-term closure only occurred in hospitals affected by the 1994 Northridge earthquake when there was structural damage; this only affected some pre-1973 hospitals. While structural damage can cause severe financial losses, the more important loss of ability to serve the community during the hours following the earthquake is more likely to be caused by nonstructural damage. At Holy Cross Medical Center, for example, damage to the air handling system and water damage from broken sprinklers and other piping required evacuation, but most services were restored within a week and paramedic units opened within 3 weeks (Figure 9-1). At Olive View Hospital (the replacement for the hospital damaged in the 1971 San Fernando earthquake) the structure was virtually undamaged (Figure 9-2), even though it was subject to horizontal ground accelerations approaching 1 g (g = acceleration of gravity). Broken piping and leakage, however, caused the evacuation of all patients and closure for one week.

During the 1994 Northridge earthquake, most nonstructural damage in healthcare facilities occurred to water related components. Damage

Figure 9-1   Exterior view of Holy Cross Medical Center, which was evacuated after the 1994 Northridge earthquake due to damage to the HVAC system. (photo courtesy of the Earthquake Engineering Research Institute)

Figure 9-2   Aerial view of Olive View Hospital, which sustained no structural damage during the 1994 Northridge earthquake, but was closed for a short while after the earthquake because of water leakage from broken sprinklers and waterlines. (photo courtesy of the Earthquake Engineering Research Institute)

was caused by leakage from sprinklers and domestic water and chilled water lines; water shortages were caused by lack of sufficient on-site storage. Twenty-one buildings at healthcare facilities suffered broken non-sprinkler water lines with most of the damage in small lines, less than 2-1/2 inches in diameter, for which bracing was not required by code. Sprinkler line breakage occurred at 35 buildings, all of which was caused by small unbraced branch lines.

Following the 1994 Northridge earthquake, a new state law was passed that required all hospitals that are deemed at "significant risk of collapse" to be rebuilt, retrofitted or closed by 2008, and all acute care hospitals to meet stringent safety codes by 2030. All hospital plans are to be reviewed by the Office of Statewide Health Planning and Development (OSHPD). The 1972 and 1994 hospital legislation is similar in scope to the 1933 and 1976 Field legislation enacted to protect schools, which is generally regarded to have been very successful in achieving its objectives of providing earthquake-safe schools.

9.4 PERFORMANCE EXPECTATIONS AND REQUIREMENTS

The following guidelines are suggested as seismic performance objectives for healthcare facilities:

- ○ Patients, staff and visitors within and immediately outside healthcare facilities must be protected at least to a life-safety performance level during design-level earthquake ground motions.

- ○ Safe spaces in the facility (which, depending on climatic conditions, may be outside) should be available for emergency care and triage activities within two hours of the occurrence of design-level earthquake ground motions.

- ○ Most hospital services should be available within three hours of the occurrence of design-level earthquake ground motions.

- ○ Emergency systems in the facility should remain operational after the occurrence of design-level earthquake ground motions.

- ○ The facility services and utilities should be self-sufficient for four days after the occurrence of design-level earthquake ground motions.

- ○ Patients and staff should be able to evacuate the building quickly and safely after the occurrence of design-level earthquake ground motions.

○ Emergency workers should be able to enter the building immediately after the occurrence of design-level earthquake ground motions, encountering minimum interference and danger.

○ There should be no release of hazardous substances as a result of the occurrence of design-level earthquake ground motions.

9.5   SEISMIC DESIGN ISSUES

The information in this section summarizes the characteristics of healthcare facilities, notes their relationship to achieving good seismic performance, and suggests seismic risk management solutions that should be considered.

Seismic Hazard and Site Issues

Unusual site conditions, such as a near-source location, poor soil characteristics, or other seismic hazards, may lead to lower performance than expected by the code design.

Unusual site conditions, such as a near-source location, poor soil characteristics, or other seismic hazards, may lead to lower performance than expected by the code design. If any of these other suspected conditions are geologic hazards, a geotechnical engineering consultant should conduct a site-specific study. If defects are encountered, an alternative site should be considered (if possible) or appropriate soil stabilization, foundation and structural design approaches should be employed to reduce consequences of ground motion beyond code design values, or costly damage caused by geologic or other seismic hazards (see Chapter 3 for additional information). If possible, avoid sites that lack redundant access and are vulnerable to bridge or highway closure.

Structural System Issues

Healthcare facilities are of great variety and size, encompassing all types of structure and services. Large hospitals accommodate several occupancy types. Acute care is a highly serviced short-term residential occupancy, and many diagnostic, laboratory and treatment areas require high-tech facilities and services. Service areas such as laundry, food service receiving, storage and distribution are akin to industrial functions, and administration includes typical office, communication and record-keeping functions.

Smaller healthcare facilities may encompass one or more functions such as predominantly longer residential care, or specialized treatment such as physical rehabilitation or dialysis. This functional variety influences some structural choices but the structure, as in all buildings, plays a background role in providing a safe and secure support for the facility activities. Since continued operation is a desirable performance objective, structural design beyond life safety is necessary and design for both structural integrity and drift control need special attention to provide an added level of reliability for the nonstructural components and systems.

The heavy and complex service demands of hospitals require greater floor-to-floor heights than for other buildings (such as offices) to provide more space above a suspended ceiling to accommodate the services. A number of hospitals have been designed with "interstitial" service space—a complete floor inserted above each functional floor to accommodate the services and make their initial installation and future change easier to accomplish (see Figure 9-3).

Because of their functional complexity, hospitals often have complex and irregular configurations. Broadly speaking, smaller hospitals are planned as horizontal layouts; large hospitals often have a vertical tower for the patient rooms elevated above horizontally planned floors for the diagnostic, treatment and administrative services. Emergency services are generally planned at the ground floor level with direct access for emergency vehicles. The structural design should focus on reducing configuration irregularities to the greatest extent possible and ensuring direct load paths. Framing systems need careful design to provide the great variety of spatial types necessary without introducing localized irregularities.

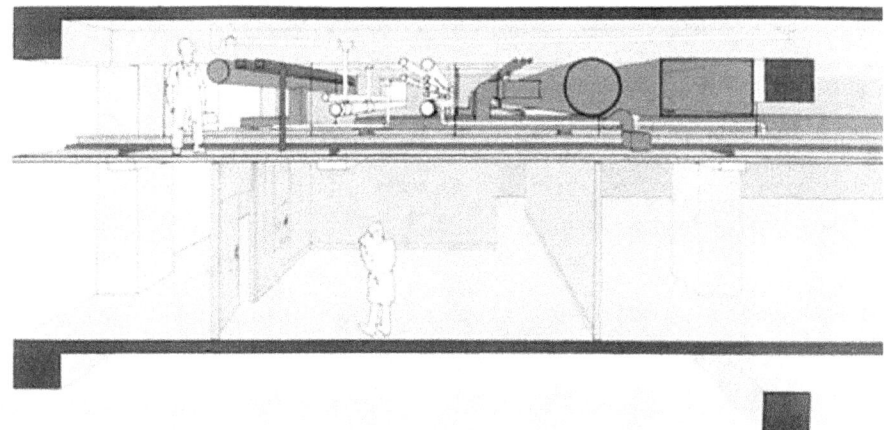

Figure 9-3   Sketch showing typical interstitial space for nonstructural components and systems in new hospitals.

As noted above excessive structural motion and drift may cause damage to ceilings, partitions, light fixtures, and glazing. In addition, storage units, library shelving, and filing cabinets may be hazardous if not braced. Excessive drift and motion may also lead to damage to roof-top equipment, and localized damage to water systems and fire suppression piping and sprinklers. Heavy equipment such as shop machinery, kilns and heavy mechanical and electrical equipment may also be displaced, and be hazards to occupants in close proximity.

Continued operation is particularly dependent on nonstructural components and systems, including purchased equipment, much of which is often of great sensitivity and cost. Many specialized utilities must be provided, some of which involve the storage of hazardous substances, such as pharmaceuticals and oxygen in tanks. These must be protected against spillage during an earthquake. Distribution systems for hazardous gases must be well supported and braced. Water must be provided to many spaces, unlike an office building, where the provision is much more limited, and thus the likelihood of water damage in healthcare facilities is greater.

The responsibilities within the design team for nonstructural component support and bracing design should be explicit and clear. The checklist for responsibility of nonstructural design in Chapter 12 (see Figure 12-5) provides a guide to establishing responsibilities for the design, installation, review and observation of all nonstructural components and systems.

10.1 INTRODUCTION

Primary and secondary (kindergarten through grade 12) schools house thousands of America's children every school day. These buildings come in a variety of configurations and sizes and are constructed from all types of structural materials like steel, concrete, masonry and wood. As school districts make decisions about the buildings that they construct, seismic considerations can easily be factored into the decision process.

The following are some unique issues associated with kindergarten through grade 12 (K-12) schools that should be kept in mind during the design and construction phase of new facilities:

○ Protection of children is an emotional societal issue and has very high priority.

○ Occupancy density is one of the highest of any building type (typically 1 person per 20 square feet by code), with the exception of summer months, and after an earthquake, children are likely to be very frightened, creating difficulties for evacuation of a damaged structure.

○ Occupancy by children is required by law, thus the moral and legal responsibilities for properly protecting the occupants are very great.

○ School facilities are critical for immediate earthquake disaster shelter and recovery efforts.

○ Closure of schools for any length of time represents a very serious community problem, and major school damage can have long-term economic and social effects.

10.2 OWNERSHIP, FINANCING, AND PROCUREMENT

Public schools are programmed and developed by the local school district. Financing is typically by local or state bond issues, possibly with the addition of federal assistance.

Public work must be competitively bid. Typically, contracts are placed for all site and building work, both structural and nonstructural. Equipment and furnishings and their installation are purchased separately from specialized vendors.

School districts typically try to emphasize high quality of design and construction and long facility life, though all districts are necessarily very budget conscious.

10.3 PERFORMANCE OF LOCAL SCHOOLS IN PAST EARTHQUAKES

There has been surprisingly little severe structural damage to schools, except in the Long Beach, California, earthquake of 1933, and there have been very few casualties. In California, no school child has been killed or seriously injured since 1933. This good fortune results primarily because all major California earthquakes since 1925 have occurred outside school hours.

Damage in the Long Beach earthquake was so severe that it was realized that if the schools had been occupied there would have been many casualties. As a result, the State passed the Field Act within a month after the earthquake. This act required that all public school buildings be designed by a California licensed architect or structural engineer, all plans must be checked by the Office of the State Architect, and construction must be continuously inspected by qualified independent inspectors retained by the local school board. The State Architect set up a special division, staffed by structural engineers, to administer the provisions of the Act. While time of day limited casualties, the Field Act, which is still enforced today, has greatly reduced structural damage.

In the Northridge, California, earthquake of 1994, State inspectors posted red UNSAFE placards on 24 school buildings, and yellow RESTRICTED USE placards on 82, although this was later considered overly conservative. No structural elements collapsed. There was, however, considerable nonstructural damage as shown in Figure 10-1. This was costly to repair, caused closure of a number of schools and, if the schools had been in session, would have caused casualties. The Field Act focused on structural design and construction, and only recently were nonstructural components included in the scope of the Act.

Figure 10-1   Nonstructural damage at Northridge Junior High where lights fell onto desks during the 1994 Northridge earthquake. (photo courtesy of the Earthquake Engineering Research Institute)

10.4   PERFORMANCE EXPECTATIONS AND REQUIREMENTS

Students and teachers within and outside elementary and secondary school buildings must be protected during an earthquake. Any damage that jeopardizes the provision of educational services impacts not only the facility but also the community, since the school is an important community center. Primary and secondary educational establishments are important community service providers and service interruption is a major problem. In addition to these general seismic performance expectations, the following guidelines are suggested as seismic performance objectives for elementary and secondary schools:

○ The school should be capable of substantial use for shelter purposes within 3 hours of the occurrence of earthquake design-level ground motions.

○ Emergency systems in the school should remain operational after the occurrence of earthquake design-level ground motions.

○ Students and teachers should be able to evacuate the school quickly and safely after the occurrence of earthquake design-level ground motions.

○ Emergency workers should be able to enter the school immediately after the occurrence of earthquake design-level ground motions, encountering minimum interference and danger.

10.5 SEISMIC DESIGN ISSUES

The information in this section summarizes the characteristics of local schools (K-12), notes their relationship to achieving good seismic performance, and suggests seismic risk management solutions that should be considered.

Seismic Hazard and Siting Issues

Unusual site conditions, such as a near-source location, poor soil characteristics, or other seismic hazards, may lead to lower performance than expected by the code design.

Unusual site conditions, such as a near-source location, poor soil characteristics, or other seismic hazards, may lead to lower performance than expected by the code design. If any of these suspected conditions are geologic hazards, a geotechnical engineering consultant should conduct a site-specific study. If defects are encountered, an alternative site should be considered (if possible) or appropriate soil stabilization, foundation and structural design approaches should be employed to reduce consequences of ground motion beyond code design values, or costly damage caused by geologic or other seismic hazards (see Chapter 3 for additional information). If possible, avoid sites that have restricted access.

Schools are a wide variety of sizes, from one-room rural school houses to 2000-student high schools. Each size will have its own code requirements and cost implications. A wide variety of structural approaches are available and careful selection must be made to meet the educational and financial program.

Traditional schools with rows of standard classrooms are relatively simple buildings, with few partitions since the structural walls can provide much of the space division. Classroom walls can act efficiently as shear walls but the school is likely to have very limited flexibility for space changes. The structure, as in all buildings, plays a background role in providing a safe and secure support for the facility activities. The structural problems are, however, relatively simple, and a well designed and constructed school should provide a safe environment.

Newer schools are usually one or two stories with light steel frame or mixed steel frame, wood and concrete or concrete masonry structures. When designed to code minimum requirements, these light and relatively long-span structures may have excessive drift characteristics. Excessive motion and drift may cause damage to ceilings, light fixtures, partitions, glazing, roof-top equipment, utilities and fire suppression piping. The structural design should pay special attention to drift control and to appropriate support of vulnerable nonstructural components and systems.

Urban schools are sometimes mid-rise (up to 4 stories), with reinforced masonry, reinforced concrete, or steel frame structures. For these structures, configuration irregularities, such as soft stories, may become critical. The structural design should focus on reducing configuration irregularities and ensuring direct load paths.

Larger schools may have long-span gymnasia or multi-use spaces in which wall-to-diaphragm connections are critical. These larger spaces may be used for post-disaster shelters. Seismic resistance must typically be provided by perimeter frames or walls. The structural design should pay special attention to reducing perimeter opening irregularities, and providing direct load path and appropriate structural connections. Larger schools also often tend to become more complex in layout because of new program needs, and the desire to provide a more supportive and attractive environment. The complexities in layout may introduce irregularities in plan shapes and require complicated framing. The structural design should focus on reducing plan irregularity, and providing appropriate structural connections.

Nonstructural System Issues

School occupants are vulnerable to nonstructural damage, particularly falling nonstructural components and systems.

School occupants are particularly vulnerable to nonstructural damage. Although school children may duck under desks and be safe from falling objects such as light fixtures and ceiling tiles, ceiling components that fall in hallways and stairs can make movement difficult, particularly if combined with power failure and loss of lighting. As discussed in the *Structural System Issues* Section, most traditional primary and elementary school buildings are relatively simple buildings, with few partitions since the structure provides the space division. Excessive motion and drift (sway) may cause damage to ceilings, partitions, light fixtures, and glazing. In addition, storage units, library shelving, and filing cabinets may be hazardous if not braced. Excessive drift and motion may also lead to damage to roof-top equipment, and localized damage to water systems and fire suppression piping and sprinklers. Heavy mechanical and electrical equipment may also be displaced.

Falling nonstructural components and systems present a significant potential for injuries to building occupants as shown in Figure 10-1. In addition to the injury potential and economic loss resulting from repair and clean-up costs, excessive service interruption can result from lighting fixture and water, mechanical, and electrical equipment damage. As discussed in the *Structural System Issues* Section, the structure should be designed for enhanced drift control to limit nonstructural damage. Lightweight hung ceilings should be avoided in light frame or large structures, and the safety of suspended lighting fixtures should always be verified. In general, the responsibilities within the design team for nonstructural component support and bracing design should be explicit and clear (Use Figure 12-5 responsibility checklist to facilitate this process).

Schools should be designed for enhanced drift control to limit nonstructural damage

DESIGN AND PERFORMANCE ISSUES RELATING TO HIGHER EDUCATION FACILITIES (UNIVERSITIES) 11

11.1 INTRODUCTION

University campuses generally consist of many different types of buildings, in a broad variety of sizes, housing many different functions. As a result, higher education facilities are, in many ways, a microcosm of the larger community. In addition to teaching classrooms, university facilities include auditoriums, laboratories, museums, stadiums and arenas, libraries and physical plant facilities, to name a few. As universities make decisions about the buildings that they construct, seismic considerations can easily be factored into the decision process.

The following are some unique issues associated with higher education facilities that should be kept in mind during the design and construction phase of new facilities:

○ Protection of students, faculty and staff is a very high priority.

○ Higher education facilities have a high daytime occupancy and some evening use, with reduced use in the summer months. Classrooms in particular often have high intensity usage.

○ Closure of higher education facilities represents a very serious problem, and major college and university damage can have long-term economic and social effects.

○ Ensuring the survival of records, whether in electronic or written form, is essential for continued operation.

○ Protection of valuable contents such as library inventories, research equipment and materials is a high priority.

11.2 OWNERSHIP, FINANCING, AND PROCUREMENT

Higher education facilities are typically developed by the institution, which may be privately, state or local-community owned. Financing of privately owned facilities is typically by private loan, possibly with some state or federal assistance; large universities also have large endowments and fund-raising activities, a large part of which assist in capital improvement program financing. Public institutions may also be financed by state and local bond issues.

Private institutions have no restrictions on methods of procurement; projects may be negotiated, conventionally bid, use construction management or design-build. Public work must be competitively bid. Typically, contracts are placed for all site and building work, both structural and nonstructural. Equipment and furnishing and their installation are purchased separately from specialized vendors.

Higher education institutions typically emphasize high quality of design and construction and long facility life, though all institutions are also budget conscious. An attractive campus is seen, particularly by institutions which are in a competitive situation, as an important asset.

11.3 PERFORMANCE OF HIGHER EDUCATION FACILITIES (UNIVERSITIES) IN PAST EARTHQUAKES

The most significant experiences of seismic performance of higher education facilities in recent earthquakes has been those related to the Whittier (Los Angeles region) earthquake of 1987, the Loma Prieta (San Francisco Bay region) earthquake of 1989, and the Northridge (Los Angeles) earthquake of 1994. During the Whittier earthquake, a number of buildings at the California State University at Los Angeles suffered some structural damage and extensive nonstructural disruption. One student was killed by a concrete facade panel that fell from a parking structure. During the Loma Prieta earthquake, the Stanford

University campus experienced considerable damage, forcing the closure of a dozen buildings. Subsequently, Stanford convened a special committee to review steps that should be taken to protect the campus against future events. One result was to set up its own seismic safety office with structural engineering staff to determine, in concert with departmental and university representatives, performance objectives for buildings and to review proposed designs. The university played a strong role in the early application of performance-based design strategies for its capital programs.

In the Northridge earthquake, the California State University at Northridge was forced to close for a month and re-open in temporary buildings. Severe damage was done to the welded steel frame of the University Library (Figure 11-1), and buildings on the University of California at Los Angeles (UCLA) campus were slightly damaged. For the most part the serious structural damage to all these campuses was experienced by older reinforced buildings or to unreinforced masonry structures.

The implications of the above-described damage caused a number of universities to become concerned about the ability of their facilities to support continued teaching and research following a more severe event.

Figure 11-1   Fractured 4-inch-thick steel base plate, university building, Northridge, 1994. (photo courtesy of the Earthquake Engineering Research Institute)

In 1997 the University of California at Berkeley committed $1 million to intensify campus planning and developed a 10-point action plan that included a high-level administrative restructuring to focus on campus planning and construction, with extensive focus on seismic safety. The 10-point plan included:

○ Creation of a new Chancellor's cabinet-level position of Vice Chancellor to oversee all aspects of the program.

○ Determination of the need for full or partial closure of any facilities deemed an unacceptable risk.

○ Development of plans for a variety of temporary relocation or "surge" space, sites and buildings.

○ Development and initiation of a multi-source financing plan to implement the master plan and implement a seismic retrofit program.

○ Conduct of a comprehensive emergency preparedness review, including mitigating nonstructural hazards, assuring that emergency and critical facilities are available, and providing emergency response training.

This plan is now being implemented; a number of key facilities have been retrofitted, and others are in process, with priorities based on a seismic evaluation of all the campus buildings. New buildings are subject to a peer-review process of the proposed seismic design.

11.4 PERFORMANCE EXPECTATIONS AND REQUIREMENTS

The following guidelines are suggested as seismic performance objectives for higher education facilities:

○ Students, faculty, staff and visitors within and immediately outside the facilities must be protected at least to a life safety performance level during design-level earthquake ground motions.

○ Emergency systems in the facilities should remain operational after the occurrence of design-level earthquake ground motions.

○ All occupants should be able to evacuate the school quickly and safely after the occurrence of design-level earthquake ground motions.

○ Emergency workers should be able to enter the facility immediately after the occurrence of design-level earthquake ground motions, encountering minimum interference and danger.

11.5 SEISMIC DESIGN ISSUES

The information in this section summarizes the characteristics of higher education facilities, notes their relationship to achieving good seismic performance, and suggests seismic risk management solutions that should be considered.

Seismic Hazard and Site Issues

Unusual site conditions, such as a near-source location, poor soil characteristics, or other seismic hazards, may lead to lower performance than expected by the code design. If any of these other suspected conditions are geologic hazards, a geotechnical engineering consultant should conduct a site-specific study. If defects are encountered, an alternative site should be considered (if possible) or appropriate soil stabilization, foundation and structural design approaches should be employed to reduce consequences of ground motion beyond code design values, or costly damage caused by geologic or other seismic hazards (see Chapter 3 for additional information). If possible, avoid sites that lack redundant access and are vulnerable to bridge or highway closure.

Unusual site conditions, such as a near-source location, poor soil characteristics, or other seismic hazards, may lead to lower performance than expected by the code design.

Structural System Issues

Higher education facilities are of great variety and size, encompassing all types of structure and services. The basic occupancies are teaching, research and administration, but assembly facilities may range from a small rehearsal theater to a multi-thousand seat sports stadium. A large student center may be a cross between a small shopping mall and a community center with retail stores, food service and places of recreation and assembly. As universities become more competitive to attract a wider audience, student-life facilities are tending to become larger and more complex. In addition, many universities provide extensive dormitory facilities.

Teaching requires spaces for small seminar groups, classrooms that are often larger in size than those of a grade school, and large lecture halls with sloped seating and advanced audio-visual equipment. Science teaching requires laboratories and support spaces with services and equipment related to traditional scientific and engineering fields, such as chemistry, biology, physics and computer sciences.

The administration function includes all office functions, including extensive communication services and extensive record keeping. Science research requires laboratories and other special facilities (e.g., greenhouses) that can accommodate a variety of unique spatial, service and utility needs required by researchers; some laboratories such as material sciences, physics, and engineering require heavy equipment with large power demands. Departmental buildings in the humanities may encompass a small administrative function, a variety of teaching facilities, many of them small. Departmental buildings in the sciences may include laboratories and their support space within the same building, and faculty offices may include direct access to research laboratories. Departmental buildings may also include a departmental library. Teaching and research in the biological sciences may include the storage, distribution and use of hazardous substances.

The library is a major campus facility, and a large campus may have several campus-wide libraries. Notwithstanding the rapid advance of computerized information technology and information sources such as the internet, the hard-copy resources of the library continue to be of major importance, and the library is a distinct building type with some specific structural and service demands, such as the ability to safely accommodate heavy dead loads, and to provide a high level of electronic search and cataloging functions.

Because of their functional complexity, large higher education facilities often have complex and irregular architectural/structural configurations. In addition, the spatial variety within many higher education buildings influences some structural choices, and structural design tends to be complex in its detailed layout with a variety of spans and floor-to-floor heights. Some laboratory equipment requires a vibration free environment, which entails special structural and mechanical equipment design. The structural design should focus on reducing configuration irregularities to the greatest extent possible and ensuring direct load paths. Framing systems need careful design to provide the great variety of spatial types necessary without introducing localized irregularities.

Since continued operation is a desirable performance objective, structural design beyond life safety is necessary and design for both structural integrity and drift control need special attention to provide an added level of reliability from the nonstructural components and systems.

Since continued operation is a desirable performance objective, structural design of higher education facilities beyond life safety is necessary and design for both structural integrity and drift control need special attention to provide an added level of reliability for the nonstructural components and systems.

Nonstructural System Issues

As noted above, excessive structural motion and drift may cause damage to ceilings, partitions, light fixtures, and glazing. In addition, storage units, library shelving, and filing cabinets may be hazardous if not braced. Excessive drift and motion may also lead to damage to roof-top equipment, and to localized damage to water systems and fire suppression piping and sprinklers. Heavy laboratory equipment and heavy mechanical and electrical equipment may also be displaced, and be hazards to occupants in close proximity.

Continued operation is particularly dependent on nonstructural components and systems, including purchased scientific equipment, much of which is often of great sensitivity and cost. Many specialized utilities must be provided, some of which involve the storage of hazardous substances. These must be protected against spillage during an earthquake. Distribution systems for hazardous gases must be well supported and braced. Water must be provided to many spaces, and thus the likelihood of water damage is greater. Cosmetic wall and ceiling damage that can easily be cleaned up in an office building may shut down a research laboratory.

> Continued operation is particularly dependent on nonstructural components and systems. Laboratory and research areas may need special design attention to nonstructural components and systems to ensure continued operation of critical experiments and equipment.

Laboratory and research areas may need special design attention to nonstructural components and systems to ensure continued operation of critical experiments and equipment.

The responsibilities within the design team for nonstructural component support and bracing design should be explicit and clear. The checklist for responsibility of nonstructural design in Chapter 12 (see Figure 12-5) provides a guide to establishing responsibilities for the design, installation, review and observation of all nonstructural components and systems.

RESPONSIBILITIES FOR SEISMIC CONSIDERATIONS WITHIN THE DESIGN TEAM 12

12.1 RESPONSIBILITIES OF THE STRUCTURAL ENGINEER, ARCHITECT, AND MEP ENGINEER

Seismic considerations should apply to every building system, sub-system, and component, and the performance of each component or system is often interdependent. The traditional organization of the design team and the assignment of responsibilities to the architect, structural engineer, MEP (mechanical, electrical, and plumbing) consultants, and other specialty consultants (e.g., geotechnical engineer, curtain wall consultant, elevator consultant, or security consultant) is critically important to address cross-cutting seismic design issues or problems.

For example, the seismic design and performance of glazing systems, windows, and curtain walls have improved significantly in recent years through the adoption of improved code provisions for these building systems. These improvements can impact both life safety in an earthquake (broken glass can kill or seriously injure) and immediate occupancy following an earthquake (integrity of the building envelope). The trade-offs involve drift limits, curtain wall clearances and design details, and glazing design. In this example, the architect, structural engineer, and curtain wall consultant must work together closely to arrive at the appropriate designs.

12.2 DEVELOPING A UNIFIED APPROACH WITHIN THE DESIGN TEAM

The first step in the design process should be the development, with active participation of the owner, of a set of clear performance objectives that address how the building is expected to perform before, during, and following an earthquake. These performance objectives should be based on owner needs and decisions, and should be expanded into detailed performance statements that apply to every sub-system of the building. Throughout the design development, there should be explicit reviews of each element of the design against the performance statements in order to assure that the completed building meets the expectations articulated in the original performance objectives. In addition, the owner should be encouraged to develop and carry out a risk management plan compatible with the performance objectives.

The term "performance objective," discussed in Chapters 2 and 4, should include a statement regarding the seismic performance that is expected of the building, subsystem, or component that is being addressed. Wherever possible, it should include quantifiable performance criteria that can be measured. For example, an objective may be that a subsystem (such as the HVAC system) should be operable following an earthquake of a certain magnitude. The specific criteria related to this may specify how long the system is expected to operate, under what operating conditions, and with what resulting interior environmental conditions.

12.3 ENGINEERING SERVICES FOR ADDED VALUE OF RISK MANAGEMENT

The owner should establish a process in which the risk management function and the facilities management function are fully coordinated in the development of a capital improvement and new construction program. The risk manager should balance seismic risk with all other

The risk manager should balance seismic risk with all other facility-related risks.

facility-related risks. In order to do so, the risk manager should have an understanding of seismic risks. Once the risk manager gains such an understanding, the risk manager should be educated to prepare a return-on-investment analysis for investments in seismic performance.

The design team has an opportunity to offer the owner a service of educating the risk manager on the details of seismic risk in buildings. This service could be independent of any specific capital improvement or design project, or it can be offered as a pre-design orientation activity that is linked to a design project.

12.4 COMMUNICATING SEISMIC CONSIDERATIONS ISSUES TO THE BUILDING OWNER

Issues of building performance should be communicated to a building owner in terms that relate how the building is expected to perform following an earthquake, and the potential impacts that this level of performance may have on the post-earthquake functionality of the building.

Issues of building performance should be communicated to a building owner in terms that relate how the building is expected to perform following an earthquake, and the potential impacts that this level of performance may have on the postearthquake functionality of the building. In order to accomplish this, the design team must learn to communicate using terminology that is familiar to the owner. This can best be accomplished through interaction with the owner's facilities or risk manager.

It is typically more difficult to explain earthquake risk issues to a building owner, since such considerations are probabilistic in nature, and less specific with respect to magnitude, location, or even how often they will occur. The design team must understand the owner's extent of risk aversion or risk tolerance. The more risk neutral the owner is, the simpler the communication is likely to be, in that various outcomes can be multiplied by their respective probabilities and then communicated directly to the owner. This process, however, becomes more complicated with a more risk averse or tolerant owner. The best way this communication can be accomplished is through close interaction and coordination with the owner's risk or facilities manager.

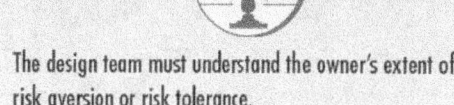

The design team must understand the owner's extent of risk aversion or risk tolerance.

As the member of the design team who initiates the design concept and develops it through design development and the preparation of construction documentation, the architect should play a key role in the seismic design process. To ensure that consideration of seismic issues occurs with the right degree of priority, and at the right time in the design process, the architect should have a clear conceptual understanding of seismic design issues that impact the design.

The structural engineer's role is to provide the structural design for a building. While the structural engineer must play the major role in providing an earthquake-resistant design, the overall design responsibility is shared between the architect and engineer, because of architectural decisions that may impact the effectiveness of the engineer's design solution and hence the building's seismic performance. The use of performance-based design can reinforce the importance of the recommendation that the architect and structural engineer work together from the inception of a design project, and to discuss seismic issues before and during the conceptual design stage. Many of the critical architectural decisions occur at the conceptual design stage, at which point the building configuration is set and issues such as the nature of the structure and structural materials and architectural finishes are identified.

The concept of structural engineers participating with architects during the early conceptual design phase of a project is not new, yet it is often confined to a cursory conversation or does not occur at all, for a variety of economic, cultural, and professional reasons. Developmental projects often require a partial design in order to procure project financing; at this point, the owner typically attempts to minimize up-front costs and the architect will not involve, or only peripherally involve, structural consultants. Some architects see the structural engineer as providing a purely service role in enabling the architect to achieve the forms and spaces that are desired. In a successful project, the architect and structural engineer typically collaborate on layout and design issues from the inception of the project, in order to ensure that the architectural and structural objectives are achieved.

DESIGN CONSIDERATION

In a successful project, the architect and structural engineer typically collaborate on layout and design issues from the inception of the project.

As the servicing needs of contemporary buildings continue to increase, the impact of the MEP (mechanical, electrical, and plumbing systems) consultant's work on seismic design becomes increasingly important. An example of this is the need for penetrations or blockouts in the structure to accommodate ductwork, piping, and equipment, which requires early design consideration. These penetrations are fundamental to the integration of the structural and mechanical system, and their size and location should be carefully worked out between the architect, structural, and mechanical engineers. There are many instance of damage to buildings in earthquakes caused by structural member penetrations that have not been adequately coordinated with the structural design.

DESIGN CONSIDERATION

There are many instance of damage to buildings in earthquakes caused by structural member penetrations that have not been adequately coordinated with the structural design.

DESIGN CONSIDERATION

Protecting against nonstructural damage requires clear allocation of roles and responsibilities.

Protecting against nonstructural damage requires clear allocation of roles and responsibilities. An important question is: Is the structural design of mechanical equipment supports the responsibility of the equipment vendor, the mechanical engineer, or the structural engineer? Similarly, is the design of the connections for precast concrete cladding the responsibility of the precast element vendor or the building structural engineer? And, is the layout and design of bracing for ductwork the responsibility of the mechanical contractor or the building structural engineer? If these responsibilities are not called out at the outset of the job, the result will be disputes, extra costs, and potentially serious omissions.

Design-Build and Fast-Track Projects

Large projects are often "fast-tracked" to some degree, with the construction contract separated into a number of bid packages that may be sole-source negotiated or competitively bid. The objective here is to speed the project's overall completion, but the process can substantially complicate coordination of tasks. Among the reasons for this are the following.

○ The complete design team may not be in existence before the preparation of construction documents has begun. This arrangement can create problems when decisions early in the project determine design approaches and delegate responsibility to entities who are not yet under contract, or who have had no input into such early decisions.

○ Communication among designers during fast-track projects is usually more difficult because the development of separate bidding packages means that the design process is fragmented, rather than one which undergoes continuous evolution. At any stage during design development and contract document preparation stages of a project, a complete set of drawings of the project may not exist.

○ Because of demands in the project schedule, the design and fabrication, or preparation of shop drawings, many items are not always thoroughly reviewed by the architect or engineer, and in some cases may not even be submitted to the local building department.

Design-build and fast-track construction can be very efficient for simple projects and for design teams that have a track record in working together, but for more complex projects and for design teams that have not previously worked together, both the design and construction phases of a project will need special attention. The assignment of roles and responsibilities is critical if the performance objectives are to be adequately defined and for integrated seismic design and construction to be achieved.

DESIGN CONSIDERATION

Design-build and fast-track construction can be very efficient for simple projects and for design teams that have a track record in working together, but for more complex projects and for design teams that have not previously worked together, both the design and construction phases of a project will need special attention.

Checklists to Facilitate the Design and Construction Process

A useful aid for the development of performance objectives and the coordination of the design and construction process within the design team is the use of checklists. These may be maintained by hand for smaller jobs, or computerized for larger or more complicated ones. Checklists can highlight key seismic design issues that require consideration and resolution, and can serve to ensure that all issues are adequately dealt with. The checklists discussed below are suggested as models that may be modified to suit the nature of the design team and the construction delivery process.

Figure 12-1 provides a seismic performance checklist, intended to focus the building owner and the design team on issues related to seismic performance expectations. The checklist presents a set of questions that are used to help the client focus on available seismic performance alternatives, leading to a recorded statement of the client's expectations of seismic performance goals that, hopefully, are in line with available

SEISMIC EXPECTATIONS

A. Earthquake Performance of Structure

Seismic Shaking Hazard Level	Damage			
	No Life Threat, Collapse	Repairable Damage: Evacuation	Repairable Damage: no Evacuation	No Significant Damage
Low				
Moderate				
High				

B. Earthquake Performance of Nonstructural Components

Seismic Shaking Hazard Level	Damage			
	No Life Threat, Collapse	Repairable Damage: Evacuation	Repairable Damage: no Evacuation	No Significant Damage
Low				
Moderate				
High				

C. Function Continuance: Structural/Nonstructural

Seismic Shaking Hazard Level	Time to Reoccupy			
	6 Months +	To 3 Months	To 2 weeks	Immediate
Low				
Moderate				
High				

Notes: Seismic Shaking Hazard Level	Spectral Acceleration (short period or 0.2 sec)	Spectral Acceleration (long period or 1.0 sec)
Low	<0.167 g	<0.067 g
Moderate	$\geq$0.167 g and <0.50 g	$\geq$0.067 g and <0.2 g
High	$\geq$0.50 g	$\geq$0.20 g

Figure 12-1   Checklist for seismic expectations. (adapted from Elssesser, 1992))

resources. Agreement on such goals and expectations forms the beginning of a performance-based design procedure and can limit future "surprises" due to unanticipated earthquake damage. The checklist statements can become a part of the project's building program, in a manner similar to statements about acoustical or thermal performance, and can serve as the basis for the use of more formal performance-based design procedures during the design.

Figure 12-2 provides a checklist intended to facilitate a discussion between the architect and the structural engineer on the importance of various building siting, layout, and design issues. The checklist identifies a number of issues that should be discussed and resolved by the architect and structural engineer at the early stages of a new project. The checklist should be used when a conceptual design has been prepared and transmitted to the structural engineer. The checklist is intended primarily to provoke a discussion, and is not intended to be filled in and used as a document of record. Most of the items in the checklist will need varying levels of discussion; the checklist is only intended to identify the existence of a potential problem and indicate the importance and priority, or significance, of the problem.

Figure 12-2 also ensures that all significant issues are covered, and that the architect and structural engineer have reached mutual understanding on the resolution of problems. This is the point at which the structural engineer should explain any issues that are not clear. Similarly, if planning or other constraints appear to have resulted in a questionable seismic configuration or a building with other undesirable seismic characteristics, the use of this checklist will ensure the identification of these characteristics fairly early in the design process, and should open the way to their resolution.

Figure 12-3 provides a list of structural and nonstructural components which are typically included in a building project. It is intended to define the responsibilities within the design team for various aspects of the design, and establishes the scope of work among the major consultants and suppliers. The checklist provides the basis for consultant agreements between the architect, construction manager, and specialist consultants. In most projects, costs and a competitive market tend to limit the time and money available for design. Working within a limited budget and timeframe, current practice is for architects and structural engineers to leave some design tasks to engineers employed by subcontractors and vendors (e.g., the design of precast concrete panels and their connections, prefabricated stairs, and truss assemblies). This

CHECKLIST TO FACILITATE ARCHITECT/ENGINEER INTERACTION			
Item	Minor Issue	Major Issue	Significant Issue
Goals			
Life Safety			
Damage Control			
Continued Function			
Site Characteristics			
Near Fault			
Ground Failure Possibility			
(landslide, liquefaction)			
Soft Soil (amplification, long period)			
Building Configuration			
Height			
Size Effects			
Architectural Concept			
Core Location			
Stair Locations			
Vertical Discontinuity			
Soft Story			
Set Back			
Offset Resistance Elements			
Plan Discontinuity			
Re-entrant Corner			
Eccentric Mass			
Adjacency-Pounding Possibility			
Structural System			
Dynamic Resonance			
Diaphragm Integrity			
Torsion			
Redundancy			
Deformation Compatibility			
Out-Of-Plane Vibration			
Unbalanced Resistance			
Resistance Location			
Drift/Interstory Effect			
Strong Column/Weak Beam Condition			
Structural System			
Ductility			
Inelastic Demand Constant or Degrading			
Damping			
Energy Dissipation Capacity			
Yield/Fracture Behavior			
Special System (e.g., base isolation)			
Mixed System			
Repairability			
Nonstructural Components			
Cladding, Glazing			
Deformation Compatability			
Mounting System			
Random Infill			
Ceiling Attachment			
Partition Attachment			
Rigid			
Floating			
Replaceable Partitions			
Stairs			
Rigid			
Detached			
Elevators			
MEP Equipment			
Special Equipment			
Computer/Communications Equipment			

Figure 12-2 Checklist for Architect/Engineer Interaction. (from Elssesser, 1992)

DESIGN SCOPE-OF-WORK GUIDELINES

Project: _____

Item	Activity					
	Design	Coordinate	Check	Shop Drawings	Sign/Stamp	Field Review
Foundation						
Super Structure Elements and Systems:						
_____						
_____						
_____						
_____						
_____						
Cladding						
_____						
_____						
_____						
Stairs						
Elevator						
Ceilings						
Equipment						
MEP Systems						

Key:

A = Architect

SE = Structural Engineer

MEP = Mechanical, Electrical, Plumbing Consultant

V = Vendor, Subcontractor or Manufacturer of manufactured, assembled or prefabricated components or systems

G = Geotechnical Engineer

___ = Other Specialty Consultant: _____

___ = Other Specialty Consultant: _____

Figure 12-3   Checklist for defining project responsibilities. Key professional personnel responsible for various aspects of design should be indicated in the appropriate cell of the check list (adapted from Elsesser, 1992).

checklist can be used to identify where and when these procedures will be used.

Figure 12-4 provides an example that shows how the checklist in Figure 12-3 may be completed for a representative project. This example shows a traditional design and construction process in which the architect plays the key role in design management and project coordination. The assigned responsibilities would vary depending on the nature of the project, the composition of the project team, and the proposed design and construction procedures.

Figure 12-5 provides a list of typical building non-structural components and, similar to Figure 12-2, is intended to delineate the roles and responsibilities of design team members for the design and installation of nonstructural components and systems. In current practice, this area is often unclear and important non-structural protective measures may become the subject of dispute; in some extreme cases, they may be omitted altogether. Both this checklist and that shown in Figure 12-2 are expected to play an important role in establishing the total scope of work for the various project consultants, and in ensuring that important tasks do not fall between the cracks of the various involved design and construction parties.

12.5 DESIGN AND CONSTRUCTION QUALITY ASSURANCE

Building codes require that "special inspections" be carried out for specific critical elements of a building during construction. These inspections are intended to assure that a high degree of quality has been achieved in constructing the approved design, and in the manner in which it is intended. As related to seismic design, special inspections typically apply to important construction and fabrication considerations, such as ensuring the use of pre-certified weld procedures and adequate weld quality.

Performance-based seismic design also requires specific performance from nonstructural systems and components in the building. In order to obtain the intended seismic performance in these areas, additional quality assurance activities are needed, above and beyond those typically required by code or employed on normal non-seismic construction projects. The following is a partial list of some nonstructural system components in need of special consideration or inspection.

DESIGN SCOPE-OF-WORK GUIDELINES

Project: ___Hypothetical*___

Item	Design	Coordinate	Check	Shop Drawings	Sign/Stamp	Field Review
Foundation	SE	A	G	SE	SE	A, SE
Super Structure Elements and Systems:						
Steel Frame	SE	A	SE	SE	SE	SE
Concrete Frame	SE	A	SE	SE	SE	SE
Precast or Post-Tensioned Floors	V	SE	SE	SE	V, SE	SE
Open Web Joists	V	SE	SE	SE	V, SE	SE
Cladding						
Precast, Stone	V	A, SE	SE	SE	V	A, SE
Metal	V	A	SE	A	V	A
Glass	V	A	A	A		A
Stairs	A, SE, V	A	SE	SE	V, SE	A, SE
Elevator	V	A	SE	A, SE	V	A, SE
Ceilings	A	A	SE	A	A	A
Equipment	V	A	SE	A	V, SE	A, SE
MEP Systems	MEP	A	SE	MEP MEP	MEP	

*This table represents a hypothetical project and should not be taken as a suggestion for assigning specific responsibilities, which must be uniquely established for each project.

Key:

A = Architect

SE = Structural Engineer

MEP = Mechanical, Electrical, Plumbing Consultant

V = Vendor, Subcontractor or Manufacturer of manufactured, assembled or prefabricated components or systems

G = Geotechnical Engineer

___ = Other Specialty Consultant: _____

___ = Other Specialty Consultant: _____

Figure 12-4 Example of completed checklist shown in Figure 12-3. (adapted from Elssesser, 1992)

NONSTRUCTURAL COMPONENT SEISMIC RESISTANCE RESPONSIBILITY MATRIX

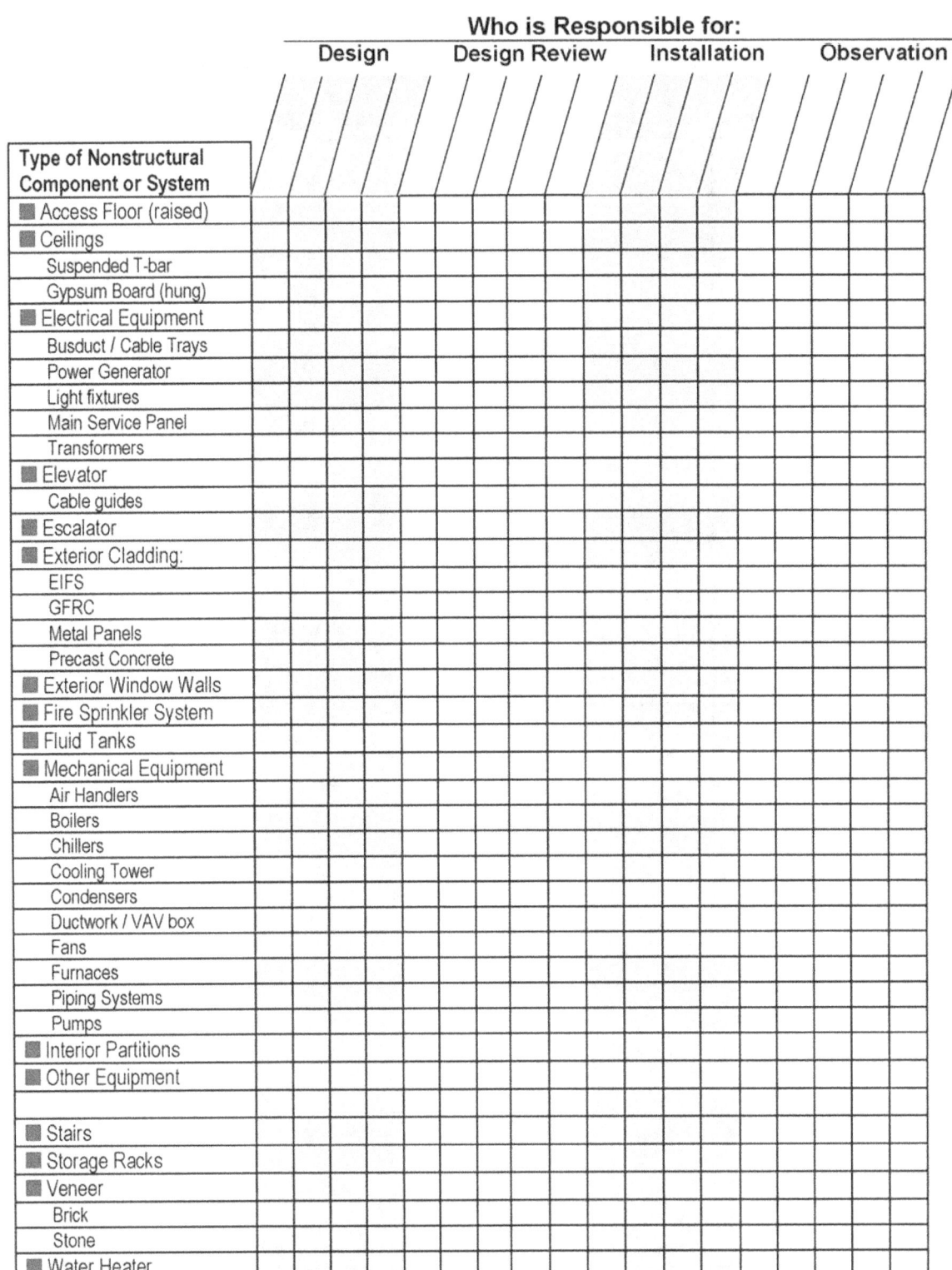

Figure 12-5 Checklist for responsibility of nonstructural component design. (from ATC/SEAOC Joint Venture, 1999)

○ Inspection of the anchorage and bracing of architectural and mechanical elements.

○ Labeling of fenestration products to ensure that they have been provided as specified, and inspection to ensure proper installation.

○ Inspection of ceiling and partition attachments.

○ Inspection of special equipment.

The report, *ATC-48, Built to Resist Earthquakes: The Path to Quality Seismic Design and Construction* (ATC/SEAOC, 1999), provides comprehensive guidance on issues pertaining to the quality design and construction of wood-frame, concrete, and masonry buildings, and anchorage and bracing of non-structural components.

Design and Construction Quality Assurance

ATC-48, Built to Resist Earthquakes: The Path to Quality Seismic Design and Construction (ATC/SEAOC, 1999).

RESPONSIBILITIES FOR SEISMIC CONSIDERATIONS WITHIN THE DESIGN TEAM

ASCE, 2002, *Minimum Design Loads for Buildings and Other Structures*, American Society of Civil Engineers, Standard *ASCE-7*, Reston, Virginia.

ASCE, 2000, *Prestandard and Commentary for the Seismic Rehabilitation of Buildings*, prepared by the American Society of Civil Engineers, published by the Federal Emergency Management Agency, *FEMA 356 Report*, Washington, DC.

ATC, 1978, *Tentative Provisions for the Development of Seismic Regulations for Buildings*, Applied Technology Council, *ATC-03 Report*, Redwood City, California.

ATC, 1989a, *Procedures for Postearthquake Safety Evaluation of Buildings*, Applied Technology Council, *ATC-20 Report*, Redwood City, California.

ATC, 1989b, *Field Manual: Postearthquake Safety Evaluation of Buildings*, Applied Technology Council, *ATC-20-1 Report*, Redwood City, California.

ATC, 1993, *ATC-20 Postearthquake Safety Evaluation of Buildings Training Slide Set*, Applied Technology Council, *ATC-20-T Report*, Redwood City, California.

ATC, 1995, *Addendum to the ATC-20 Procedures for Postearthquake Safety Evaluation of Buildings*, Applied Technology Council, *ATC-20-2 Report*, Redwood City, California.

ATC, 1996a, *Case Studies in Rapid Postearthquake Safety Evaluation of Buildings*, Applied Technology Council, *ATC-20-3 Report*, Redwood City, California.

ATC, 1996b, *Seismic Evaluation and Retrofit of Concrete Buildings*, Volumes 1 and 2, Applied Technology Council, *ATC-40 Report*, Redwood City, California.

ATC, 1997, *Example Applications of the NEHRP Guidelines for the Seismic Rehabilitation of Buildings*, prepared by the Applied Technology Council for the Building Seismic Safety Council; published by the Federal Emergency Management Agency, *FEMA 276 Report*, Washington, D.C.

ATC, 2002, *Recommended U.S.-Italy Collaborative Procedures for Earthquake Emergency Response Planning for Hospitals in Italy*, Applied Technology Council, *ATC-51-1 Report*, Redwood City, California.

ATC, 2004a (in preparation), *Improvement of Inelastic Seismic Analysis Procedures*, prepared by the Applied Technology Council for the Federal Emergency Management Agency, FEMA 440, Washington, D.C.

ATC, 2004b (in preparation), *Program Plan for Development of Performance-Based Seismic Design Guidelines*, prepared by the Applied Technology Council for the Federal Emergency Management Agency, FEMA 445, Washington, D.C.

ATC, 2004c (in preparation), *Characterization of Seismic Performance for Buildings*, prepared by the Applied Technology Council for the Federal Emergency Management Agency, FEMA 446, Washington, D.C.

ATC/BSSC, 1997a, *NEHRP Guidelines for the Seismic Rehabilitation of Buildings*, prepared by the Applied Technology Council (ATC-33 project) for the Building Seismic Safety Council, published by the Federal Emergency Management Agency, *FEMA 273 Report*, Washington, DC.

ATC/BSSC, 1997b, *NEHRP Commentary on the Guidelines for the Seismic Rehabilitation of Buildings*, prepared by the Applied Technology Council (ATC-33 project) for the Building Seismic Safety Council, published by the Federal Emergency Management Agency, *FEMA 274 Report*, Washington, DC.

BSSC, 1988a, *Seismic Considerations – Apartment Buildings*, prepared by the Building Seismic Safety Council, published by the Federal Emergency Management Agency, *FEMA 152 Report*, Washington, DC.

BSSC, 1988b, *Seismic Considerations – Office Buildings*, prepared by the Building Seismic Safety Council, published by the Federal Emergency Management Agency, *FEMA 153 Report*, Washington, DC.

BSSC, 1988c, *NEHRP Recommended Provisions for Seismic Regulations for New Building and Other Structures, Part 1 – Provisions*, prepared by the Building Seismic Safety Council, published by the Federal Emergency Management Agency, *FEMA 152 Report*, Washington, DC.

BSSC, 1990a, *Seismic Considerations – Elementary and Secondary Schools*, prepared by the Building Seismic Safety Council, published by the Federal Emergency Management Agency, *FEMA 149 Report*, Washington, DC.

BSSC, 1990b, *Seismic Considerations – Health Care Facilities*, prepared by the Building Seismic Safety Council, published by the Federal Emergency Management Agency, *FEMA 150 Report*, Washington, DC.

BSSC, 1990c, *Seismic Considerations – Hotels and Motels*, prepared by the Building Seismic Safety Council, published by the Federal Emergency Management Agency, *FEMA 151 Report*, Washington, DC.

BSSC, 1995, *NEHRP Recommended Provisions for Seismic Regulations for New Buildings*, 1994 edition, prepared by the Building Seismic Safety Council, published by the Federal Emergency Management Agency, *FEMA 222A Report*, Washington, DC.

BSSC, 1998, *NEHRP Recommended Provisions for Seismic Regulations for New Buildings and Other Structures*, 1997 edition, prepared by the Building Seismic Safety Council, published by the Federal Emergency Management Agency, *FEMA 303 Report*, Washington, DC.

BSSC, 2001, *NEHRP Recommended Provisions for Seismic Regulations for New Buildings and Other Structures, Part I: Provisions*, and *Part II, Commentary*, 2000 Edition, prepared by the Building Seismic Safety Council, published by the Federal Emergency Management Agency, Publications, *FEMA 368 Report* and *FEMA 369 Report*, Washington, DC.

Benuska, L. (Editor), 1990, *Loma Prieta Earthquake of October 17, 1989: Reconnaissance Report*, Supplement to *Earthquake Spectra*, Volume 6, Earthquake Engineering Research Institute, Oakland, California, 450 pages.

Comartin, C.D. (Editor), 1995, *The Guam Earthquake of August 8, 1993: Reconnaissance Report*, Supplement to *Earthquake Spectra*, Volume 11, Earthquake Engineering Research Institute, Oakland, California, 175 pages.

Comartin, C.D., Greene, M., and Tubbesing, S.K., (Editors), 1995, *The Hyogo-ken Nanbu Earthquake, January 17, 1995: Preliminary Reconnaissance Report*, Earthquake Engineering Research Institute, Oakland, California, 116 pages.

Chung, R. (Editor), 1995, *Hokkaido-nansei-oki, Japan, Earthquake of July 12, 1993: Reconnaissance Report*, Supplement to *Earthquake Spectra*, Volume 11, Earthquake Engineering Research Institute, Oakland, California, 166 pages.

Cole, E.E. and Shea, G.H., (Editors), 1991, *Costa Rica Earthquake of April 22, 1991: Reconnaissance Report*, Special Supplement B to *Earthquake*

Spectra, Volume 7, Earthquake Engineering Research Institute, Oakland, California, 170 pages.

EERI, 1985, *Impressions of the Guerrero-Michoacan, Mexico, Earthquake of 19 September 1985: A Preliminary Reconnaissance Report*, Earthquake Engineering Research Institute, published in cooperation with the National Research Council of the National Academy of Sciences, Oakland, California, 36 pages.

EERI, 2000, *Action Plan for Performance-Based Seismic Design*, prepared by the Earthquake Engineering Research Institute, published by the Federal Emergency Management Agency, *FEMA 349 Report*, Washington, DC.

Elsesser, E., 1992, "Structures for Seismic Resistance," *Buildings at Risk : Seismic Design Basics for Practicing Architects*, AIA/ACSA Council on Architectural Research, Washington, DC.

Frankel, A.D., and Leyendecker, E.V., 2000, "Uniform hazard response spectra and seismic hazard curves for the United States: CD-ROM," U.S. Geological Survey, National Seismic Hazard Mapping Project.

Frankel, A., Mueller, C., Barnhard, T., Perkins, D., Leyendecker, E., Dickman, N., Hanson, S., and Hopper M., 1996, *National Seismic Hazard Maps: Documentation June 1996*, U.S. Geological Survey, Open-File Report 96-532, 110 p.

Frankel, A.D., Mueller, C.S., Barnhard, T.P., Leyendecker, E.V., Wesson, R.L., Harmsen, S.C., Klein, F.W., Perkins, D.M., Dickman, N.C., Hanson, S.L., and Hopper, M.G., 2000, "USGS national seismic hazard maps," *Earthquake Spectra*, V. 16, No. 1, Earthquake Engineering Research Institute, Oakland, California, pages 1-19.

Hall, J. (Editor), 1995, *Northridge Earthquake of January 17, 1994: Reconnaissance Report*, Vol. I, Supplement to *Earthquake Spectra*, Volume 11, Earthquake Engineering Research Institute, Oakland, California, 523 pages.

Hall, J. (Editor), 1996, *Northridge Earthquake of January 17, 1994: Reconnaissance Report*, Vol. II, Earthquake Engineering Research Institute, Oakland, California, 280 pages.

ICBO, 1997, *Uniform Building Code*, International Conference of Building Officials, Whittier, California.

ICC, 2000, *International Building Code*, International Code Council (formerly: Building Officials and Code Administrators International,

Inc., International Conference of Building Officials, and Southern Building Code Congress International, Inc.), Birmingham, Alabama.

Malley, J.O., Shea, G.H., and Gulkan, P., (Editors), 1993, *Erzincan, Turkey, Earthquake of March 13, 1992: Reconnaissance Report*, Supplement to *Earthquake Spectra*, Volume 9, Earthquake Engineering Research Institute, Oakland, California, 210 pages.

NFPA, 2003, *5000 Building Construction and Safety Code*, National Fire Protection Association, Quincy, Massachusetts.

SAC, 2000a, *Recommended Seismic Design Criteria for New Steel Moment-Frame Buildings*, prepared by the SAC Joint Venture, a partnership of the Structural Engineers Association of California, Applied Technology Council, and California Universities for Research in Earthquake Engineering, published by the Federal Emergency Management Agency, *FEMA 350 Report*, Washington, DC

SAC, 2000b, *Recommended Specifications and Quality Assurance Guidelines for Steel Moment-Frame Construction for Seismic Applications*, prepared by the SAC Joint Venture, a partnership of the Structural Engineers Association of California, Applied Technology Council, and California Universities for Research in Earthquake Engineering, published by the Federal Emergency Management Agency, *FEMA 353 Report*, Washington, DC

SBCCI, 1997, *Standard Building Code*, Southern Building Code Congress, International, Birmingham, Alabama.

Schiff, A.J. (Editor), 1991, *Philippines Earthquake of July 16, 1990: Reconnaissance Report*, Special Supplement A to *Earthquake Spectra*, Volume 7, Earthquake Engineering Research Institute, Oakland, California, 150 pages.

SEAOC, 1995, *Vision 2000: Performance-Based Seismic Engineering of Buildings*, Structural Engineers Association of California, Sacramento, California.

SEAOC, 1999, *Recommended Lateral Force Requirements and Commentary*, prepared by the Structural Engineers Association of California, published by the International Conference of Building Officials, Whittier, California.

USACE, 1998, *Seismic Design for Buildings*, U.S. Army Corps of Engineers, Publication TI809-04.

Working Group on California Earthquake Probabilities, 2003, *Earthquake probabilities in the San Francisco Bay region: 2002-2031*, U. S. Geological Survey, Open-file Report 03-214, Menlo Park, California

Wyllie, L.A. and Filson, J.R., (Editors), 1989, *Armenia Earthquake of December 7, 1988: Reconnaissance Report*, Supplement to *Earthquake Spectra*, Volume 5, Earthquake Engineering Research Institute, Oakland, California, 175 pages.

Ziony, J. I. (Editor), 1985, *Evaluating Earthquake Hazards in the Los Angeles Region—An Earth Science Perspective*, U. S. Geological Survey, Professional Paper 1360, Reston, Virginia.

Principal Investigator

Christopher Rojahn
Applied Technology Council
201 Redwood Shores Parkway, Suite 240
Redwood City, California 94065

FEMA Project Officer

Milagros Kennett
Federal Emergency Management Agency
500 "C" Street, SW, Room 416
Washington, DC 20472

Senior Review Panel

John Ashelin
Project Engineer - Corporate Engineering
Anheuser-Busch Companies, Inc.
One Busch Place
St. Louis, Missouri 63118

Deborah B. Beck
Beck Creative Strategies LLC
531 Main Street - Suite 313
New York, New York 10044

Clifford J. Carey AIA
University of Illinois
Project Planning & Facility Management
807 S. Wright Street, Suite 320
Champaign, Illinois 61820

Edwin T. Dean[1]
Nishkian Dean
319 SW Washington Street, Suite 720
Portland, Oregon 97204

---

1.   ATC Board Contact

Randal C. Haslam AIA CSI
Director of School Construction
Jordan School District
9150 South 500 West
Sandy, Utah 84070

Roger Richter
California Healthcare Association
1215 K Street, Suite 800
Sacramento, California 95814

Project Working Group

Chris Arnold
Building Systems Development
P.O. Box 51950
Palo Alto, California 94303

Craig D. Comartin
Comartin-Reis Engineering
7683 Andrea Avenue
Stockton, California 95207-1705

David B. Hattis
Building Technology Inc.
1109 Spring Street
Silver Spring, Maryland 20910

Patrick J. Lama
Mason Industries
350 Rebro Drive
Hauppauge, New York 11788

Maurice S. Power
Geomatrix Consultants, Inc.
2101 Webster Street, 12th Floor
Oakland, California 94612

Evan Reis
Comartin-Reis Engineering
345 California Ave, Suite 5
Palo Alto, California 94306

www.ingramcontent.com/pod-product-compliance
Lightning Source LLC
Chambersburg PA
CBHW081445170526
45166CB00008B/2323